D0786443

Flowers and Plants

Flowers and Plants

An International Lexicon with Biographical Notes

Robert Shosteck

Foreword by James L. Crowe,
Assistant Director,
U.S. Botanic Garden

Quadrangle/The New York Times Book Co.

Contents

Foreword

Here is a publishing event all naturalists—amateur and expert alike—can give thanks for. At last we have a book that surveys several thousand wild and cultivated plants and alphabetically arranges each plant according to its common name.

Concentrating on flora found in North America and Canada, *Flowers and Plants* provides the scientific name of each plant, followed by an accurate and descriptive account of its origin, use, and history. Such interesting facts as food value, medicinal content, dye possibilities, fiber content, and religious aspects and folklore uses are included in each description and make reading Mr. Shosteck's book a joy. What a wealth of information to have at one's fingertips, whether for trying to answer questions on plants identified only by their common names or just plain browsing.

The reader can identify more readily hundreds of species, or differentiate among closely related species or genera, by an understanding this book provides of the meaning of Latin and Greek scientific names.

One learns, for example, that the wild cucumber's name, *Echinocystis,* which means "hedgehog" and "bladder," refers to the prickly, bladder-like fruit. If interested in gardening, it is useful to know, for example, that Baby's Breath's name, *Gypsophila,* means "to love chalk," in reference to the shrub's preference for a lime of chalky soil.

The reader will note the frequent occurrence of certain specific names, all useful in recognizing a plant. Examples are: *trifolium,* "three-leaved," *umbellata,* "umbrella-like, *convolvulus,* "twining around," and *bidens,* "two-toothed (seeds)."

A plant's edibility is denoted by such species names as *esculenta,* "edible," *oleracea,* "like a potherb," and *hortense,* "of the garden." Its usefulness in the marketplace or in the apothecary's is denoted by the species name, *"officinalis."* Also, *arvensis,* "of cultivated fields," signifies a plant's economic significance.

This volume offers the reader an acquaintance with notable figures in botanical and horticultural history, from Pliny and Linnaeus to those of the nineteenth and twentieth centuries. The author provides biographical sketches of the great plant explorers and collectors, herbalists, hybridizers, gardeners, botanical writers and artists, who are remembered by the scientific plant names honoring them—like zinnia, fuchsia, dahlia, and begonia.

Perusing the book, the reader can gain a new insight into the origins and relationships of various plants. One learns why the pumpkin, squash, and gourd are closely allied; why such diverse vegetables as cabbage, kale, kohlrabi, broccoli, brussels sprouts, and turnips all stem from a common ancestor.

It may come as a surprise to many to learn of the role that women have played in botanical history, as evidenced by the many plants named in honor of women. The earliest was Zenobia, the third-century queen of Palmyra. Other women notables honored by flower names are: the Fitton sisters in England, Mrs. C. Van Brunt of the United States, the Dutchess of Northumberland, the noted Kew Gardens artist, Matilda Smith, and Queen Maria Louisa Theresa.

Of added value are the several hundred line drawings, which should prove useful as an aid in identifying many of the plants listed in the book.

Every horticulturist, botanist, landscape architect, nurseryman, florist, home gardener and house plant lover should have this handy and fascinating excursion through the greenness of the world.

James L. Crowe,
Assistant Director,
U.S. Botanic Garden

Acknowledgments

The author expresses his gratitude to several individuals who were most helpful in reviewing and commenting on the Introduction, and in advising on the many problems of nomenclature and significance of scientific names. Among these were James Crowe, Assistant Director, U.S. Botanic Gardens, Washington, D.C.; Stanwyn G. Shetler, Associate Curator, Department of Botany, Smithsonian Institution, Washington, D.C.; and Dr. Edward E. Terrell, botanist, National Agricultural Research Center, Beltsville, Maryland.

He is also grateful to the reference staff at the National Agricultural Library, Beltsville, Maryland, for the many hours devoted to assisting him in locating elusive journals and books, and in suggesting reference works useful in his research.

The marginal illustrations used in this book came from a number of sources. The author is most appreciative of the generosity of the publishers or authors of the following books for permission granted to use illustrations from their works.

Carolina Landscape Plants, by Professor R. Gordon Halfacre, North Carolina State University, Raleigh, N.C.
Flora of West Virginia, by Professor Earl L. Core, University of West Virginia, Morgantown, W. Va.
Michigan Wildflowers, by Helen V. Smith, Cranbrook Institute of Science, Bloomfield Hills, Mich.
Native Plants of Pennsylvania, Bowmans Hill State Wildflower Preserve, Washington Crossing State Park, Pa.
Selected Weeds of the United States, Agricultural Research Service, U.S. Department of Agriculture, Washington, D.C.
Wildflowers of Ohio, by I. H. Klein, Cleveland Museum of Natural History, Cleveland, Ohio.

Acknowledgments

Woody Plants of Maryland, by Dr. R. G. Brown, University of Maryland, and Melvin L. Brown, Frostburg State College, Maryland.

I would like to thank Edward S. Gruson, author of *Words for Birds,* which inspired me to do a lexicon on my love, botany.

Introduction

Mankind has given names to plants since the beginning of time. Those useful to him as food and animal fodder, medicine or decorative purposes came first. Plants used in primitive religious rites or in witchcraft, or which were abundant and colorful, likewise got names. These often reflected their uses, habitats or notable characteristics.

These vernacular names remained part of a long oral tradition, extending for thousands of years. Not until the invention of printing were plant names stabilized with respect to their spelling and to specific kinds of plants to which they applied.

We are indebted to the early Greeks for the impetus to the infant science of botany. Theophrastus (372–287 B.C.) recorded almost five hundred plant names, and Pliny the Elder (A.D. 23–79), described about a thousand plants in his *Natural History*. Dioscorides, active in the first century, described about six hundred plants, mostly medical species.

Most plants remained nameless until some time after 1500, when interest began to develop in natural history in Europe. This led to the beginning of botany as a science, and with it the study, description and naming of hitherto nameless plants.

Since our focus is on English plant names, we will attempt to trace their origins. A large number go back to Anglo-Saxon words whose origins are lost to us. Many names were brought to Britain by the Romans, and also with the introduction of Christianity. This was true of plants of culinary or medical use, and also of those with religious associations. These were classical names of Latin or Greek origin, which were taken over into English. Some were corrupted or abbreviated so that their origins are recognizable only to a person with an etymological background. Such common names as lettuce, parsley, lily, mallow and milfoil all are rooted in classical languages.

A number of plant names came into English as a consequence of the early development of European trade with the Far East. Plants of economic value, especially spices, aromatics, medicinals, dye and fiber plants, were brought by traders to Rome and Greece. Trade in these plant products later spread through most of Europe. As a trading nation, modern England has had commercial intercourse with nations and peoples on all of the continents. These contacts brought into the English language plant names from the principal tongues of the world. Often, these became so abbreviated or modified in the process that one familiar with the original language would scarcely recognize the word. Nevertheless, these assimilations greatly enriched our language. A few examples follow: German, thistle; French, pansy; Spanish, yucca; Italian, artichoke; Arabic and Hebrew, jasmine; Irish, shamrock; Norse, madder; Persian, tulip; Greek, clematis; Hebrew, fig.

A great many English plant names originated during the Middle Ages, and gained currency through the herbals published from the sixteenth through the eighteenth centuries. In the absence of common names, the English herbalists did not hesitate to coin or to adapt names, often making translations into English of Latin, Greek or German plant names. The vernacular names which survived usually were simple words and easily remembered through association with some noteworthy characteristic of the plant concerned. A tear-thumb was not an easily forgotten name if its stem prickles tore one's thumb. The lizard's tail really resembles the tail of a lizard. Similarly, the sensitive plant's leaves fold on being touched; the waxy, yellow buttercup appears to be a miniature cup of butter; the globeflower looks like a globe.

American Indian languages contributed scores of plant names to English. Early settlers learned from the aborigines of the medicinal virtues, real or imagined, of many native herbs, with the resultant squawroot, Indian physic and colicroot. They also learned of many edible plants from the Indians, such as pokeweed, persimmon, squash and hickory, incorporating these names into the English language.

Simultaneous with the adoption of these Indian names, the early settlers in America bestowed Old World plant names on native plants which outwardly resembled plants of their former homeland. Thus the primroses and cowslips which are not at all related to their European namesakes.

The Doctrine of Signatures, in vogue in the seventeenth and eighteenth centuries, held that the medicinal virtues of a plant were denoted by something in the appearance of the plant that resembled the disease symptoms, or afflicted organ it was supposed to cure. It was based upon the belief that God, in this way, had indicated its special virtues.

Coles *Art of Simpling* states explicitly the origin of the "Doctrine of Signatures." He writes, "The mercy of God which is over all his Workes, maketh Grasse to grow upon the Mountaines, and Herbes for the use of Man, and hath not only stamped upon them a distinct forme, but also has given them particular Signatures, whereby a man may read, even in legible characters, the use of them." Thus, if a child suffered from nettle rash, nettle tea was given as a cure. Turmeric, used in making a yellow dye and in curry powder, was used as a remedy for jaundice, because of the yellow color of the skin, characteristic of that disease. Similarly, the spotted leaves of pulmonaria made that plant a remedy for tuberculosis. Hepatica, or liver-leaf, was believed to cure liver ailments, because of the liver-shaped leaf.

This doctrine, largely discarded by the close of the eighteenth century, resulted in the naming of many common plants for the reputed curative properties they possessed.

Natural history had an important place in the derivation of vernacular plant names. Hundreds of such names are based upon the real or fancied resemblance of flowers, flower parts, leaves, or seeds of the plants; or to animals, birds, reptiles, fish or insects; or to organs or parts of any of these living forms. Thus we find lamb's-quarters, coltsfoot, dandelion, foxglove and cattail; also, columbine, cranesbill, lizard's tail and dragonhead.

Another group of names alludes to the use of plants as food by various birds, animals, or other creatures. Edible properties account for goosegrass, duckweed, chickweed and pigweed; also catnip, bee-balm and partridgeberry. A group of plants bear compound names beginning with such words as *horse* and *bull*. These usually denote the coarse, wild or unpalatable character of the plant so designated. Thus we find horsemint, bullrush, horse nettle, horseradish, horseweed and horse chestnut.

In some cases animal names are used in a derogatory sense to distinguish the inferior or false species from the superior or true species. There are, for example, toadflax and horse chestnut, which are not true flaxes or chestnuts.

Snakes have an important place in plant names because of their role in mythology and folklore, and the fact that the venomous species long have been feared. Thus, there are at least four species known as snakeroot because of supposed curative properties for snakebite. There are also snakemouth, snaketongue, addersmouth and rattlesnake-master.

Another group of names with the ending "bane" originated because of the real or fancied capacity of the plants to either repel or injure the given animal. Among familiar examples are wolfbane, cowbane, henbane, and fleabane. Bane, of course, is related to the word banish.

Several score names of common plants have their origin in analogies to mankind, his everyday objects and customs. Certain resemblances led to lady's slipper, jack-in-the-pulpit, buttonbush, shepherd's purse, prince's feather and fiddlehead. More fanciful are such names as virgin's bower, Venus' looking-glass and angel's trumpet.

Also related to human activities are plant names based upon their use in arts and crafts. Examples are teasel, madder, dyeweed and wild indigo. Supposed medicinal virtues of certain plants account for many names. Among these are statice, bugloss, squill, boneset, feverfew, heal-all, scabious, wormseed, lungwort, and milkwort.

Scores of names are derived from the food use of

plant species. All sorts of berries, nuts, beans, and other fruits and roots have been named with distinguishing prefixes. Berries run the gamut from bearberry to whortleberry. There are dozens of specific kinds of nuts and roots. Other names relating to food use include salad burnet and meadow parsnip.

Many plants received their designations through chance religious associations. St. Johnswort flowers on that saint's day, while the resemblance of the St. Andrew's Cross to its namesake led to that name. The origin of the Star of Bethlehem should be obvious to all. Passionflower received its designation because of the resemblance of colors and number of flower parts to the elements in Christ's passion. At least a dozen saints are remembered through flower names.

Notable personalities in botanical history account for a large number of common names. In virtually all cases these were adopted from the Latinized generic names. Among well-known flowers in this category are: zinnia, dahlia, fuchsia, begonia and forsythia. There are probably thousands of generic names that recall notable personalities. These will be discussed later.

Roman and Greek mythology have contributed a large number of plant names to English. In most instances these are presently used generic names. Some examples are iris, anemone, centaurea, crocus, narcissus, gentian, and violet. Many eighteenth- and nineteenth-century botanists, and especially Linnaeus, were well-informed in classical mythology. Whenever an opportunity presented itself they used a mythological name to designate a new genus, though it often required flights of fancy to relate some of the plant's characteristics to the mythological character. Dozens of these names were accepted as common names in the absence of vernacular alternatives.

Through the centuries those interested in plant life observed that some flowers preferred or grew only in certain habitats; among rocks and stones; in marshes, bogs or swamps; along the seacoast; in woodlands; in ponds or sluggish streams; or in open fields. These distinctive habitats gave rise to vernacular plant names, such as woodsorrel, bog asphodel, stonecrop, rockcress and pondweed.

Another group of plants received their designations because of the season in which they flower or when their fruit or seeds are conspicuous. Thus we have spring beauty, summer savory, autumn crocus and winterberry.

Thousands of plants, especially those from tropical habitats, never have had common names known to the civilized community. If vernacular names have existed, they have not been recorded. Many of these plants which have gained popularity have become identified by their scientific names, which in time became their common names.

Confusion regarding plant names arose at an early date as different common names were given to the same species in different countries or regions within a country. The same name was applied to similar or even quite dissimilar species.

Large-scale immigration to America in the eighteenth and nineteenth centuries helped to compound the confusion in plant names. Hundreds of European species were brought to our shores, often in ballast or as an adulterant in seed or in grain. Often these received new vernacular names in America when they became commonplace weeds.

The language barrier in plant names was keenly noted by the eighteenth-century botanists and horticulturists, who, as a consequence, used Latin plant names in their communications and publications.

Carolus Linnaeus (1707–1778), the noted Swedish botanist, put the botanical house in order when he established a system of binomial nomenclature. This bestowed a generic and a specific name on each kind of plant, assigning related genera to plant families (or "natural orders," as Linnaeus called them). In his landmark work, *Species Plantarum* (1753), Linnaeus named and described 5,900 species, assigning each to a genus and family. Today, about 300,-000 different plant species have been described and categorized according to his system.

Linnaeus studied at Lund and at Uppsala universities in Sweden. At the age of 23 he was appointed lecturer in botany at Uppsala. Three notable works brought him lasting fame. These were *Systema Naturae* (1735), the first systematic classification of the animal world; *Genera Plantarum,* an enumeration of plant genera, and *Species Plantarum.*

Linnaeus's classification was based on the sexual characteristics of plants as the determinants of natural affinities. Later botanists broadened the basis of classification to include genetics, form, structure and cell studies. The families into which plants are grouped are recognized by "eae" or "ae" endings, as Violaceae, for the violet family. It is helpful in recognizing plants to learn the key characteristics common to members of each family. Similarly, when one makes the acquaintance of many of the species in a given genus or family, he gains a conception of the great changes wrought by the evolutionary process, especially in plant adaptations to environmental changes, and to the role of man in altering cultivated plants to suit his needs.

The International Code of Botanical Nomenclature, which developed much later, is recognized worldwide as the guide to establishing scientific names of plants.

In America, horticulturists, gardeners, herbalists and wildflower enthusiasts ordinarily use the English common names, with which they are most familiar. They can always fall back on the Latin botanical name to make certain of the identity of the plant to which they are referring.

In the present work, every plant is listed alphabetically by its common English name, followed by its scientific or botanical name. The latter consists of the generic name, followed by species name or epithet: A genus (derived from the Latin word for "race"), is made up of a group of closely related species, with certain common attributes. The generic name, a noun, is always capitalized, while the species name, never capitalized, is either an adjective, or a noun used adjectivally. An example will serve to clarify the relationship.

The generic name for clover is *Trifolium,* meaning "three-leaved." Certain species in this genus are for the most part readily differentiated by the form and color of the flowers, as white, red or yellow clovers. Similarly, the violets botanically are *Viola*. There are blue, cream, yellow, white and purple violets, even two-colored ones. Other violet species are separated by marked differences in leaf form.

Linnaeus, his contemporaries and successors, used many different bases for forming generic names. They made use of ancient classical names used by the Greeks and Romans; their physicians, philosophers, medical and botanical writers. Many of these names originated in mythology. Often, these ancient generic names have no relationship to the group of plants to which the names were latterly assigned.

Hundreds of generic names commemorate the name of a botanist, plant explorer, naturalist, missionary, gardener, patron of horticulture, a plant's discoverer or a botanical artist. Such names are Latinized with a suitable ending, such as *Kalmia*, from the name Kalm, and *Zinnia*, from the name Zinn.

A large number of vernacular names from different languages have been Latinized and accepted as botanical names. This has given such local names international standing and precise meaning.

Generic names often are descriptive, alluding to the color or form of the flower, leaves or seeds; or to the plant's habitat or place of origin. Other names refer to an unusual or striking trait of the plant; or to an attribute common to all members of the genus. Typically, Latin or Greek words were used in creating these descriptive generic names.

The specific name or epithet owes its origin to much the same circumstances as the generic name. These can be descriptive, as *flavus*, meaning yellow; geographic, as *virginiana;* or personal, as *lewisii*, for Meriwether Lewis, American explorer.

A brief discussion of the spelling and pronunciation of scientific names may be helpful to the reader. Latin generic names are nouns and have the appropriate gender endings. Specific names are generally Latin adjectives which agree in gender with the genus. There are all sorts of exceptions to the rules, however.

Regarding pronunciation, the accent is on the

first syllable of a two-syllable word such as canna. Accent next to the last on more than two-syllable words when the vowel is long, as in pilosum. In words of three or more syllables, if the next to the last syllable has a short vowel, it is accented, e.g., grandiflora.

Botanical names have a precise, recorded origin and history. The botanist who names and classifies a new plant places it in its appropriate family and genus, and then publishes a full botanical description of the new species in a botanical journal. He may construct a descriptive name from Latin or Greek words, or name it in honor of someone he admires. The botanist's name is forever associated with the new species by an abbreviation or initials after the name. Best known is L. for Linnaeus.

Illustrations often accompany the original description, and the botanist also preserves in a herbarium a type specimen. This is dried and mounted on a sheet of stiff paper. Should a dispute arise, the specimen can be examined to determine if it is really different in important details from earlier-described related species. Botanists engaged in taxonomic work (i.e., classification) usually have access to a large herbarium, often connected with a university or scientific institution, which they can use in comparing specimens to determine whether a new species really has been found.

Occasionally, one will find a plant with a generally recognized common name, listed in two different works under differing generic and/or scientific names. This indicates that the process of scientific plant-naming is not fool-proof. Differences of opinion as to the proper botanical names for certain genera or species have always existed, and will continue to be a problem of minor proportions despite the detailed rules established under the International Code of Botanical Nomenclature. One basic problem is the varying interpretation to the significance of differences in species. One botanist may say that these differences are only the natural variations to be found within a species, while another will proclaim two distinct species. These differences are usually resolved over the years by consensus, or by more detailed study.

Horticultural varieties, or cultivars, with very few exceptions, are considered beyond the scope of this work. The work of plant breeders over the centuries has resulted in the naming of tens of thousands of horticultural varieties, hybrids and mutants of cultivated ornamentals and vegetables. These names are often registered with some national or international authority. The exact rules for naming such plants are embodied in the code of nomenclature established in the International Code of Nomenclature of Cultivated Plants, 1969.

Cultivar names are distinguished from the botanical binomials and varieties since they are not printed in italics, but are enclosed in single quotes or preceded by the abbreviation "c.v." Hybrids are distinguished by an X sign between the two parental names, or by the use of the word "hybrid" in the name.

Horticultural varieties are occasionally mentioned in this work, especially in reference to some common vegetables. The variety may be quite distinctive outwardly, but it is often the creature of man, or a selection from the wild. An example may be helpful. *Brassica oleracea* var. *capitata* is the botanical name for cabbage. *B. oleracea* was a European seacoast plant with a thick, hard stem and yellow flowers. Long cultivation and selection has resulted in many interesting varieties within this species. Kale closely resembles the original wild type. From the wild prototype man has produced cauliflower, broccoli, kohlrabi and brussels sprouts.

Flowers and Plants

Abelia to Azalea

Abelia *Abelia graebneriana*

Two noted contributors to botany are commemorated in this generic name. Clark Abel was a physician at the embassy of Lord Amherst in Peking, China, in the early 1800s. An amateur botanist when his medical duties were not pressing, he collected extensively and recorded his work in *A Narrative of a Journey in the Interior of China . . . 1816–1817*. His large collection of dried specimens was lost in the wreck of the *Alceste* en route to England.

Carl O. Graebner (1871–1933) began his career as an assistant in the Botanic Garden in Berlin in 1891. By 1910 he was the head of University of Berlin Botanic Garden as well as a professor.

A. Graebneriana is the taller of two popular species. The smaller is *A. Engleriana,* also named for a noted German botanist.

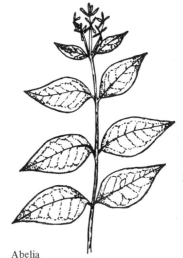

Abelia

Achimenes *Achimenes spp.*

This tender houseplant cannot stand the slightest chilling, as we learn from its Greek name, *a cheimino,* "not to suffer from cold." About forty species, many of them hybrids, are available in a wide range of colors. The popular gloxinia is a close relative.

Actinidia, Bower *Actinidia arguta*

Two technical floral characteristics account for the scientific names. The generic name is derived from the Greek word for "ray," in reference to the radiate styles, and *arguta,* signifying "sharp toothed," alludes to the serrate leaf margins.

The bower portion of the name describes a common use of this vine as a cover or bower for garden shelters. The large greenish-yellow berry is edible.

1

Green Adder's Mouth

Adder's Mouth, Green *Malaxis unifolia*
Malaxis, Greek for "a softening," alludes to the tender tissues of this plant; *unifolia* is Latin for "one leaf," which is all a plant bears. The small flowers of this inconspicuous orchid are suggestive of an adder's open mouth.

Adder's tongue. See Trout lily

Adonis, Summer *Adonis aestivalis*
This plant was named for the youthful hunter of Greek mythology who was beloved by Aphrodite. The Latin word *aestivalis* refers to its summer-flowering habit. This attractive garden annual bears scarlet flowers.

Aechmea *Aechmea fasciata*
This epiphyte, or air plant, from tropical America derives its name from the Greek *aichme,* meaning "a point," which refers to the stiff tips of the sepals on the flowers. *Fasciata,* meaning "bound together," refers to its floral characteristics.

African Violet *Saintpaulia spp.*
This violet-like flower from East Africa was named in honor of Baron Walter von St. Paul-Illàire (1860–1910), who discovered it. A German military officer and colonial administrator in German East Africa, he was strongly interested in botany.

Ageratum, Annual *Ageratum conyzoides*
Ageratum, derived from two Greek words meaning "not old," has flowers that retain their bright colors for a long time and do not age rapidly. This suggests a new word in English for someone who doesn't show his age.
Conyzoides means "resembling Conyza," the plowman's spikenard. This blue-flowered annual is very popular as a border plant.

Agrimony *Agrimonia eupatoria*

Once spelled argemoney, this name derives from the Greek *argemos,* a "white speck on the eye," an affliction cured with an extract made by boiling agrimony. The specific name honors Eupator, a king of Pontus (near the Black Sea), who in his spare time was something of a herbalist. Health-giving properties have been attributed to agrimony tea for a long time. Pour a pint of boiling water over several tablespoons of dried leaves, let cool, then strain and drink.

Akebia, Five-leaf *Akebia quinata*

Akebia is the Latinization of the Japanese vernacular name for this twining vine. Its five leaflets, like the fingers of a hand, account for the species name. *Akebia* is noted for its semi-evergreen, graceful foliage and large edible pods.

Alfalfa *Medicago sativa*

Medicago stems from the Greek *medike,* "from Medea," the supposed country of origin of this widely distributed forage plant. *Sativa,* Latin for "sown as a crop," attests to its value as animal feed, as does the common name which is derived from Arabic and signifies "best fodder." Alfalfa was introduced from Arabia into Italy and then into Spain. It finally reached England in the seventeenth century.

Alfalfa contains 14 of 16 essential minerals and Vitamins A, D, and K. The young, tender leaves and flower tips, edible in spring and early summer, may be used in a salad or in cole slaw or with cooked cereal. The dried leaves can be used as a substitute for India tea. Pulverized alfalfa is an ingredient in a popular baby food.

Agrimony

Allamanda *Allamanda cathartica*

These shrubs and vines from Brazil were named after Dr. J. N. S. Allamand of Leyden, Holland, who presented seeds of this species to the famed

Carolus Linnaeus, the father of modern botany. The allamandas have yellow or purplish flowers and a prickly fruit. Its specific name reflects its medicinal properties.

Allegheny Vine. See Climbing Fumitory

Aloe *Aloe spp.*
Aloe is the original Arabic name for these perennial succulent herbs, of which there are more than 150 species. Most bear fleshy, stiff, spiny leaves in rosettes. These are ideal indoor pot plants.

Aluminum Plant *Pilea cadieri*
Pilea, derived from the Latin word meaning "a Roman cap," refers to the shape of the flower. *Cadieri* honors a botanist and plant collector. The quilted pattern leaves with silvery-gray markings suggested the common name of this house plant.

Amaranth, Green *Amaranthus hybridus*
Amaranth, derived from the Greek word for "unfading," refers to the lasting quality of the plant in the fall. The specific name stems from the belief that this species originated as a natural hybrid.

Young plants are used as potherbs and have been described as equal to spinach in flavor and palatability. The numerous tiny black seeds are useful in baking. Gather the ripe tops in plastic bags before the seeds ripen. Use a roller to "thresh" the seed, and blow away the chaff.

Amaryllis (hybrid) *Hippeastrum spp.*
This is the popular amaryllis sold for early spring bloom. The generic name, made up of two Greek words meaning "knight" and "star," is based on the supposed resemblance of these flowers to the related Barbados lily.

This flower was dedicated in ancient days to Amaryllis, a shepherdess in Greek mythology. The word actually means "sparkling," which alludes to the attractiveness of these flowers.

Amaryllis *Amaryllis belladonna*
Belladonna, meaning "beautiful lady," was originally applied to a medicinal plant. This amaryllis originated in the Cape of Good Hope area and has been much improved by hybridization. A true lily, the flowers bloom in triplets, and flower parts also come in threes.

Anchusa, Alkanet *Anchusa italica*
This plant is important in the history of cosmetics since its Greek name means "paint for the skin." Fashion-conscious ladies of ancient Greece used a rouge made from its roots. Anchusa juice was used to color wine, leather, and lip salves. Alkanet stems from the Spanish *alcana* and the Arabic *al-henna,* meaning henna dye. The dye, however, originally came from an Egyptian shrub.
Italy is the land of origin of this blue garden perennial.

Andromeda, Mountain *Pieris floribunda*
Mt. Pierus in Thessaly, Greece, was sacred to the Muses. It was chosen for the generic name of this plant by a classically-minded botanist. The species name is descriptive of the numerous small, waxy-white flowers borne in terminal clusters or spikes in early spring.
The common name, also of classical origin, is an earlier generic name. Andromeda was the daughter of Cepheus and Cassiopeia, rulers of Ethiopia. According to Greek mythology, she was chained to a rock as an offering to a sea monster. She was rescued by Perseus, who married her.
This broad-leaved evergreen, related to the azalea, is native to the Appalachian Mountains.

Angelica, Garden *Angelica officinalis*
According to an old legend, the curative virtues of

this herb were revealed to a monk by an angel during a terrible plague, hence the name. At one time the dried leaves were believed to protect against contagious diseases, purify the blood, and even cure alcoholism. The species name refers to its place in the apothecary shop as a medical herb.

In Europe the leaves and stalks are blanched and eaten as celery. Angelica often is found in gardens of European-born Americans. A member of the parsley family, angelica has a strong scent, bears greenish-white flowers, and has finely divided leaves. The following species entry identifies the edible parts of the stalk and roots.

Angelica, Wild *Angelica atropurpurea*

See *Angelica, Garden,* for derivation of the generic and common names.

Atropurpurea, Latin for "dark purple," is descriptive of the purple-splotched stem of this species, found in wet bottomlands and swamps. The stems are used in making angelica preserves. Root-stocks, young shoots, and stems can be candied. Boil them thoroughly for about fifteen minutes, discard the water, and boil them again in heavy sugar syrup. Then cool and dry. As a cooked vegetable, angelica is strongly suggestive of celery.

Wild Angelica

Angel's Trumpet *Datura arborea*

The generic name traces back to the old Arabic word *tatorali. Arborea,* or "tree-like," describes the branching habit of this poisonous plant, which is related to the thorn apple or jimsonweed.

Angel's trumpet bears white tubular flowers that grow as long as nine inches. It is an annual and is easily grown from seed. Related Asiatic and Middle Eastern species have been used in religious rites for over a thousand years. The seed causes temporary hallucinations and mental derangement.

Anglepod *Gonolobus obliquus*

The common name refers to the sharp angles of

the seed pod of one species. *Gonolobus* is a Greek-derived word meaning "angle-pod." The specific name *obliquus* refers to the unequal sides of the leaf lobes. This wild vine bears small brown-purple flowers and large heart-shaped leaves.

Anise *Pimpinella anisum*

Anisum is the old Latin name of pimpernel. The latter, as *pimpinella,* a Middle Latin word, became the generic name for a group of aromatics, including anise and dill.

Anise has long been used in baking and as a flavor in soups, liquors, and confections. Anise tea is made by pouring half a pint of boiling water over two tablespoons of mashed seeds. The dried seeds have a long history of medicinal use as a purgative and a cure for flatulence, stomach ache, and liver and kidney ailments.

In the fall potato-like tubers are produced at the ends of underground runners. Boiling or roasting makes them sweetish and palatable.

Aphelandra *Aphelandra squarrosa*

This tropical American plant with variegated leaves and handsome red flowers has gained some popularity as a house plant. Its Greek name, meaning "simple man," does not refer to mankind. It signifies that the flowers have simple, one-celled anthers, or male parts, of the flower. The specific name indicates that the leaves are spread widely from the axis or stem.

Apple-of-Peru *Nicandra physalodes*

Nicander of Colophon, a Greek poet who also wrote about plants and their medicinal uses, is commemorated by this generic name. Its specific name alludes to its resemblance to *Physalis,* a related genus.

The common name refers to the large, bladderlike husk and the plant's Peruvian origin. This is strictly an ornamental plant; the berries are poisonous.

Trailing Arbutus

Arbutus, Trailing *Epigaea repens*

The creeping habit of this early spring wildflower is the basis for its generic name, Greek for "upon the ground." The Latin species name also refers to its creeping. The Latin word *arbutus* refers to the European strawberry tree (no kin to our strawberry), which bears a tiny seed capsule resembling that of the trailing arbutus. This sweet-scented woodland denizen is the state flower of Massachusetts and the floral emblem of Nova Scotia.

Arnica *Arnica mollis*

Two sources present different accounts of the origin of *arnica*. We offer both without comment. One states that the name is a Latin corruption of *ptarmica*, from the Greek *ptarmikos*, "causing to sneeze." This refers to supposed properties of arnica. The other version derives the name from the Greek *arnakis*, "a lambskin," alluding to the soft texture of its leaves. *Mollis*, Latin for "soft-hairy" is descriptive of the leaf texture and supports the second account.

The roots of this plant have long been used in pharmacy as a stimulant and for local irritant effect.

Arrowhead *Sagittaria latifolia*

Sagittaria, Latin for "arrow," describes the shape of the leaves, and *latifolia* translates to "broad leaves." The common name is based on the Latin name. This denizen of pond and stream banks bears white flowers in threes.

Artichoke, Globe *Cynara scolymus*

A Greek derivative, the word *kyon*, "dog," refers to the guard leaves around the flowers, which resemble dog's teeth. *Scolymus* is Latin for "artichoke." This is the thistle-like plant grown for its edible flower heads. A native of the Mediterranean, it is grown in the Southern half of the United States. The common name is from the Italian *articiocco*, the vernacular name for this plant.

8

Artillery Plant *Pilea microphylla*
See Aluminum Plant for the origin of the generic name.

Microphylla, "small-leaved," distinguishes this from related species.

The common name arose because the stamens violently discharge, or "shoot out," their pollen when dry and mature (a sort of botanical orgasm). This species is also grown for its compact, fern-like sprays.

Asparagus, Garden *Asparagus officinalis*
This Latin word was borrowed from the Greek *asparagos,* "a sprout or shoot," and refers to the food use of the tender sprouts. *Officinalis* is the Latin word for the usefulness of this plant as food or medicine.

Asparagus frequently escapes from cultivation and can be found along rural roadsides. There are separate male and female plants, the latter bearing the berries which are spread by birds excreting the seeds. Close observation will reveal that each male flower has a rudimentary ovary; each female flower, a rudimentary stamen.

Asparagus Fern *Asparagus plumosus*
See Asparagus, Garden, for the derivation of the generic name.

Plumosus is Latin for "feathery" and alludes the appearance of the finely divided leaves. This decorative house plant is used widely by florists.

Aspidistra *Aspidistra lurida variegata*
One must closely examine the flower of the aspidistra to discover that the stigma is shaped like a "small, round shield," the meaning of the Greek generic name. The "lurid" purple flowers account for the species name, and the green and white-striped leaves explain the varietal name. This popular house plant, with stiff, shiny foliage, withstands adverse conditions, such as excessive heat, cold, or drought.

Aspidistra

9

Aster, China *Callistephus chinensis*

"A very beautiful crown" is the translation of the Greek generic name, an apt designation for this popular garden annual. Many have three-inch double flowers. *Chinensis,* "from China," alludes to the country of origin.

Its well-known common name derives from the Latin and Greek word for star. According to an old Greek legend, the aster was created out of star dust when Virgo, looking down from heaven, wept bitterly. Sacred to all the gods, wreaths of asters were placed on temple altars on all festive occasions.

New York Aster

Aster, New York *Aster novae-belgii*

See also Aster, China.

The specific name, meaning "New Belgium," reveals the classical thinking of Linnaeus, who bestowed this name, and his emphasis on the original or primary.

Linnaeus knew that the Dutch had first settled the area that later became New York. Searching for a classical synonym for New Netherlands, he found that the original Roman province which included modern Holland was named Belgica. Thus it was logical to name this the New Belgium aster.

This aster was introduced into England over a century ago. Through hybridization and selection British horticulturists developed the Michaelmas daisy, now a repatriate well-known to American gardeners.

Aster, White-topped *Sericocarpus asteroides*

Sericocarpus, derived from the Greek, means "silk-fruited." This refers to the silky hairs covering the seeds. *Asteroides,* "resembling an aster," suggests the appearance of this plant. The common name alludes to the conspicuous white bracts surrounding each flower head.

Astilbe; False Goatsbeard *Astilbe simplicifolia*

This Greek word, meaning "not shining," alludes to the dull leaves and insignificant individual flowers of this plant. *Simplicifolia* means "simple-leaved," that is, entire or undivided leaves.

False goatsbeard indicates this plant resembles but is unrelated to the goatsbeard. The latter name, a very old one, arose from the resemblance of the dense spikes of white flowers to a goat's beard.

Aucuba *Aucuba japonica*

The name of this well-known ornamental plant is the Latinization of the vernacular Japanese name, "Aokiba." *Japonica* notes its Japanese origin.

Aucuba enjoys popularity because of its large evergreen variegated foliage and its bright scarlet fruit. It is hardy as far north as Washington, D.C., and grows in sheltered locations farther north.

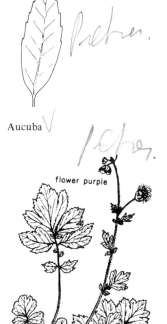

Aucuba

flower purple

Avens, Purple *Geum rivale*

Geum, the classical name of this herb, is perpetuated as a generic name. *Rivale,* Latin for "pertaining to brooks," refers to its preferred habitat. *Avens* is derived from Latin *avencia,* a kind of clover, and then from Old French, *avence.* A well-boiled avens root makes a chocolate-like drink that is slightly acidic and astringent. The addition of sugar makes it more palatable.

Avocado *Persea americana*

Avocado sprang from the ancient Aztec word *ahuacotl,* which means "testicle tree." The shape suggested this name to the imaginative Aztec priests, who also imputed aphrodisiacal virtues to the luscious fruit. The Spaniards corrupted the Aztec name to *aquacate.* This was altered by the English, much later, to *avocado.*

Purple Avens

Later botanists, ignorant of the origin of the avocado, named it *Persea,* under the erroneous notion that it had originated in Persia. The specific name, *americana,* was a belated effort to rectify the error by indicating its origin in the Americas.

Home gardeners often sprout the large egg-shaped seed, which grows into a graceful slender sapling that usually survives for only a few years.

Pink Azalea

Azalea, Pink *Rhododendron nudiflorum*

Rhododendron, Greek for "rose tree," alludes to the color effect of the masses of flowers. The specific name, the Latin for "naked flower," may refer to the fact that the flowers appear long before the leaves. They are naked, without leaves on the twigs. *Azalea* derives from Greek, meaning "dry," and refers to the wild shrub's habitat in dry, rocky woods.

It is interesting to compare the wild azalea with the hundreds of cultivated varieties and to note the great improvements in color, number, and size of flowers effected through breeding and selection.

Baby's Breath to Buttonbush

Baby's Breath *Gypsophila repens*
The generic name, which translated literally means "love chalk," refers to the plant's preference for chalky soil. *Repens,* "creeping," is indicative of the growth habit of the species. Also popular is G. *elegans,* a truly "elegant" flower. The common name was bestowed because the flowers' fragrance resembles an infant's breath.

Baby Tears *Helxine soleirolii*
Helxine is from the Greek word meaning "to tear." The plant was so named because the seeds readily catch onto the clothes of passersby. Joseph F. Soleirol (died 1863) built up a large collection of Corsican plant species in the mid-1800s. This small creeping herb with a mass of tiny leaves is suitable for alpine gardens. Baby tears is the fanciful name given this plant because of its streamers of tiny leaves.

Bachelor's Button *Centaurea cyanus*
Centaurea, recalling the mythological monster, was the classical name for this plant. *Cyanus,* Greek for "blue," refers to its typically deep blue flowers, though several varieties have white or pink flowers. The English, practical even in naming flowers, thought that the flower heads resembled buttons, hence bachelor's buttons.

Balloon-flower; Japanese Bellflower *Platycodon grandiflorum*
The Greek generic name, translated "broad bell,"

is aptly descriptive of the appearance of these bell-flowers. *Grandiflorum,* "large-flowered" in Latin, informs us that this species is distinguished from others by the size of its flowers. The buds resemble small balloons, hence the common name.

Balloon Vine *Cardiospermum halicacabum*
The black seeds of this cultivated vine have heart-shaped spots, a unique feature which accounts for the generic name, "heart-seed" in Greek. *Halicacabum* is a classical name applied to this species for reasons not now known. The bladderlike capsules accounts for the common name of this vine.

Balsam, Alpine *Erinus alpinus*
The Greek name refers to the early flowering habit of this alpine plant. A supposed resemblance to balsam accounts for the common name. This rosy purple perennial is a rock garden favorite.

Balsam, Garden *Impatiens balsamina*
The significance of the Latin-derived generic name is known to the observant gardener: When the pods are ripe, a slight touch results in a violent explosive discharge of seeds in all directions. *Balsamina* and the common name suggest the resemblance of this species to an older balsam. This species also is popularly known as touch-me-not. There are white, rose and, scarlet flowered varieties.

Balsam, Sweet or Lemon *Melissa officinalis*
The Greeks named this herb *melissa,* or "honey-bee," because its flowers were so attractive to the bees. *Officinalis* connotes that it was sold in the marketplace as a culinary or medicinal herb.
Balsam comes from the Latin *balsamum,* or "balsam tree," and earlier from the Hebrew *besem,* meaning "spice." This herb has been cultivated for over 2,000 years as an aromatic and a bee plant. It is still used today to flavor soups, stews, and sauces. To make balm tea, pour a cup of boiling water over

4 ounces dried balm leaves, steep 10 to 15 minutes, cool, and strain.

Balsam Apple; Wild Cucumber *Echinocystis lobata*

Botanists had no trouble finding a. name for this plant. The Greek word for "hedgehog," *echinos,* and for "bladder," *kystis,* describe the prickly, bladderlike fruit. The triangular lobed leaves led to the species name, *lobata.*

The fruit resembles a small cucumber, hence the common name. The other name has a historical background. Early herbalists called garden balsam *balsamina.* This was corrupted to balsam apple by unlettered English gardeners. At a later date, for reasons unknown, the name was applied to this species.

Baneberry *Actaea pachypoda*

Linnaeus bestowed this generic name, anciently that of the elder, because of somewhat similar leaves. He turned again to Greek for *pachypoda,* which means "thick foot," possibly referring again to the leaf-shape.

Baneberry warns of the baneful consequences of eating the poisonous berry.

Balsam Apple

Barberry, Japanese *Berberis thunbergi*

Berberis is the Latinized version of the Arabic name for this shrub. Barberry, in turn, is a corruption of the generic name.

Carl P. Thunberg (1743–1828) was sent by Linnaeus on a plant-collecting expedition to Japan and Java and was appointed professor of botany at Uppsala University in Sweden. Linnaeus chose this species to honor his associate.

Barberry is not grown in the vicinity of wheat fields since it is the host for one stage of the life cycle of wheat rust, a fungus disease which can be disastrous to a wheat crop.

There are over 175 species of barberry in Asia and the Americas, several of which are under cultivation.

Japanese Barberry

Bartonia

Bartonia *Bartonia virginica*

This small herb in the gentian family was named in honor of B. S. Barton, a versatile scientist, botanist, and physician (1766–1815). He wrote the first botany text in the United States (1803), which went into six editions. He wrote *Flora Virginica* (1812) and *Natural History of Pennsylvania* (1791) and was professor of natural history, botany, and materia medica at the University of Pennsylvania. He also found time to serve as curator of the collections of the American Philosophical Society (1790–1800) and as its vice-president (1802–1815). To Thomas Jefferson's dismay, his *Natural History of the Lewis and Clark Expedition* was never published. As a young man he joined David Rittenhouse in surveying the western boundary of his state (1785).

Basil. See Savory, Summer

Bastard Toadflax *Comandra umbellata*

This generic name is from the Greek, *kome* and *andros,* "hair" and "man." The botanist who bestowed this name wasn't thinking of a hairy man; rather, it was his observation that the hairy calyx lobes were attached to the anthers. *Umbellata* refers to the resemblance of the flowerhead to the framework of an umbrella.

The common name means that this plant resembles but is unrelated to the true toadflax. *Comandra* is a root parasite on other plants. It plays the game both ways since it can also manufacture food through its green leaves. The fully grown but unripe nutlets

Bastard Toadflax   are sweet and make a delicious nibble.

Bayberry *Myrica pensylvanica*

Myrica is the ancient name of an aromatic shrub. The species name denotes the state from which the first specimen was described.

The common name indicates a preferred habitat of this shrub—the sandy shores of bays and inlets along the Atlantic coast, from Maryland northward. Bayberry is almost evergreen, has aromatic foliage,

Bayberry

and bears gray, wax-covered berries. They are often collected in the fall for candle-making. It takes 1.5 quarts to make an 8-inch candle. The berries are boiled in water and the wax is skimmed off. This can be mixed with paraffin to yield more candles.

Beach Heather. See False Heather

Beach Pea *Lathyrus maritimus*
Lathyrus is the ancient Greek name for pea or pulse—later applied to this genus. It is derived from *la* and *thouros,* meaning "addition" and "irritant," which refer to the medicinal uses of the plant. *Maritimus,* "of the seashore," locates the preferred habitat of this species. The tender young peas are edible but not very palatable.

Bean, String, or Snap *Phaseolus vulgaris*
Phaseolus, the old Latin name for kidney bean, was later applied to the garden string bean. *Vulgaris,* "common," is the specific name for the wild tropical American bean that is the ancestor of many of our cultivated forms, including kidney, snap, navy, bush, and climbing beans. *Snap bean* arose from the snapping noise when a bean is broken in two. *String bean* refers to the string along the edge of the pod, though many "string beans" are virtually stringless today.

Bean, Wild *Strophostyles umbellata*
The Greek *strophos* means "a turn or twist" of a flower's style and refers to its curved appearance. The flowerhead of this pink wild bean is structured like an umbrella frame, hence the specific name *umbellata.* A closely allied species, S. *helvola* (Latin for "yellowish" or "greenish"), is the trailing wild bean. Its flowers change color upon maturing.

Bean, Yard-long *Vigna sesquipedalis*
This cultivated bean was named in honor of Dominic Vigni, a seventeenth-century Paduan commenta-

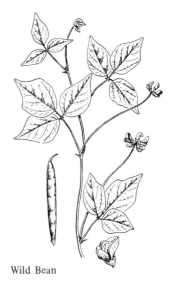

Wild Bean

tor on Theophrastus and a noted botanist. The specific name means "a foot and a half long," an ungenerous name for a bean that can easily reach a 3.5-foot length.

Bearberry *Arctostaphylos uva-ursi*
The Greek generic name signifies "bear grape," since in the past it was eaten by bears. The specific name means the same, but in Latin. The common English name reemphasizes the Latin and Greek.
If they are well cooked, the dried berries can be used as an emergency food, but they are better left for the bears.

Beardtongue *Pentstemon digitalis*
The Greek-derived generic name means "five stamens." Each flower has five stamens, of which one is always sterile. *Digitalis* is Latin, "in the form of fingers," and refers to the shape of the corolla, which resembles the foxglove. *Beardtongue* was the name given by early botanists who considered the bearded, tongue-shaped sterile stamen an oddity.

Bearberry

Bear's-foot. See Leafcup

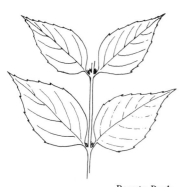

Beauty Bush *Kolkwitzia amabilis*
Richard Kolkwitz, professor of botany at Berlin University in the late nineteenth and early twentieth centuries, conducted research and wrote extensively in the field of plant physiology. He wrote *Ecology of the Brazilian Plant World* and studied alkaloids in the narcissus. The specific name, Latin for "lovely," is an apt designation, as is the common name.

Beauty Bush

Bedstraw; Catchweed *Galium aparine*
Gala, the Greek root word for "milk," recalls the widespread use of galium in curdling milk for making cheese. *Aparine,* meaning "holding" or "clinging," refers to the tiny hooked prickles found on the seeds and stems of this plant.

This plant was once used as straw for beds, as a mattress, and as tick filling, both in England and America. The leaves of this widespread genus come in whorls of four, six, or eight, some prickly, others smooth. A seventeenth-century authority, John Ray, stated that a refreshing beverage could be made by distilling the flower tops of this weed.

Bee Balm; Oswego Tea *Monarda didyma*

This genus honors Nicholas Monardes (1493–1588), a botanist and physician in Seville, Spain. He wrote a book, *Joyfull News Out of the New Founde Worlde,* which described useful American plants. An English edition was issued in 1577. The first monarda was brought from Virginia to England in 1637, where it rapidly gained favor.

The specific name means "twinned" or "paired" in Greek and refers to the pairs of stamens on each flower.

Beebalm suggests that the fragrant flowers are attractive to bees. Oswego tea, the other common name, arose from the fact that John Bartram first found this plant at Oswego, then an outpost on Lake Ontario, where it was used for tea. He wrote that "it combines fragrant leaves with brilliant flowers." Bartram sent some seed to his English friend Peter Collinson in 1744. By 1760 its popularity was such that it was "plenty in the Covent Garden Market."

Beechdrops *Epifagus virginiana*

Epifagus is Latin for "upon the beech" and refers to the beechdrop's parasitism on the roots of the beech. The branched plants, which lack green pigments, bear small yellowish to brownish flowers. *Virginiana* refers to the state from which the species was first described. The plant collector should have no problem finding this parasite within ten or fifteen feet of a beech tree.

Beefsteak Plant. See Perilla

Beet, Garden *Beta vulgaris*

Beta, from which "beet" is derived, is an ancient

Beechdrops

Latin name. *Vulgaris* means "of common occurrence." The ancestral slender-rooted species is found in sandy soil on the shores of the Mediterranean Sea. Its cultivation began about 300 years before the Christian era.

Beggar's-lice. See Stickseed

Beggar-ticks; Sticktight *Bidens frondosa*

Beggar-ticks

Bidens refers to the "two teeth," or prickles, on each seed. *Frondosa* refers to the "frond-like" appearance of the finely dissected leaves. The tiny hooked seeds easily become attached to the clothing of passersby in fields or along paths, hence the common names.

Begonia *Begonia spp.*

This popular house plant was named in honor of Michel Begon (1638–1710), patron of botany, governor of French Canada, and superintendent at Santo Domingo. Over 200 varieties and thousands of subvarieties of begonia are known to horticulturists. Plant explorers have brought home new types from Mexico, South America, Asia, and Africa. Both floral and foliage begonias have been developed. *B. rex* is best known of the latter group.

The sexes are separate. Male flowers have four petals, female flowers, five, with a winged ovary below.

Although most begonias are of ornamental interest, some tropical species have medicinal uses. They are used to reduce certain fevers and as a purgative. One species has been used in the treatment of syphilis.

Bellflower *Campanula americana*

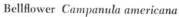

Bellflower

The generic name is a diminutive of the Latin word for "bell," *campana,* and alludes to the bell-shaped flowers. There are six or more wild species and many cultivated forms, including Canterbury bells. The fleshy rootstock makes a sweetish salad or a parsnip-like dish when cooked. *Americana* refers to the American origin of this species.

Bellflower, Japanese. See Balloon-flower

Bells-of-Ireland *Molucella laevis*

The Moluccas, or Spice Islands, in western Asia are regarded as the original habitat of this genus. The species name, meaning "smooth," refers to the leaves.

When the white corolla withers and falls, the calyx that remains looks like a little green bell, which explains the common name. This flower is a popular garden herb and is used in flower arranging.

Bellwort *Uvularia perfoliata*

This plant once was considered a remedy for diseases of the uvula, the fleshy lobe in the middle of the soft palate. A pair of opposite leaves surrounds the stem as if it were one leaf, hence the name *perfoliata* which means "through the leaf." *Bellwort* alludes to the bell-shaped flowers of this herb. *Wort* is from the Anglo-Saxon *wyrt,* meaning "root or herb."

The young shoots of this woodland denizen can be eaten like asparagus. We think it is far more interesting as a wildflower.

Bellwort

Bergamot *Monarda fistulosa*

See Bee Balm for the derivation of *Monarda.*

The specific name, from Latin, means "pipe- or reed-like" and refers to the shape of the flower tube. The common name is believed to derive from Bergamo, Italy. This species bears pink or pale lilac flowers and is sometimes seen in cultivation. Related varieties have purple or white flowers.

Bergenia, Heart-leaved *Bergenia saxifraga*

This genus honors Karl A. von Bergen (1704–1766), professor of botany at Frankfurt University.

He wrote several works on medical botany and *Flora of Frankfurt* in 1750. Bergen also held a professorship in anatomy. The king of Prussia ordered all bodies of executed criminals sent to von Bergen for classroom dissection and demonstration. *Saxifraga*, Latin for "growing in rocky crevices," describes the habitat of this plant.

Bindweed, Field or Hedge *Convolvulus sepium or arvensis*

This generic name derives from the Latin *convolvo,* "to twine around," an apt description of the twining habit of this vine. There are two widespread species. *Sepium* means "of hedges or fences" and refers to the typical habitat; *arvensis* means "of cultivated fields" and reflects that species' typical abode. Both have white or pink flowers.

Birthwort; Virginia Snakeroot *Aristolochia serpentaria*

This plant, of obstetrical interest, derives its generic name from the Greek words *aristos* and *lochia,* meaning "best childbirth." Its roots were reputed to aid in easing birth pains during labor if chewed and eaten prior to delivery. *Serpentaria* means "snakelike" and refers to the twisted appearance of the rootstock. The common name also alludes to the use of the root at childbirth.

The alternate name originated with a tribe of Indians in Virginia. They believed that the mashed roots of this herb, applied as a wet compress to a snakebite, would effect a cure. This species was once listed in the U.S. Pharmacopoeia because of its tonic and stimulant properties.

Birthwort

Bishop's-hat *Epimedium alpinum*

Epimedium is an ancient Greek plant name used by Pliny, so named because Medea was believed to have been its original habitat. Its original alpine abode is reflected in the species name. The common name is thought to relate to a fancied resemblance of the rosy purple flowers to a bishop's hat. This species is used as a ground cover in shaded areas and rock gardens.

Bishop's-weed, Mock *Ptilimnium capillaceum*
The generic name is a combination word denoting leaf form and habitat. *Ptilon* is the Greek word for "feather" and alludes to the feathery leaves; *Limno* means "marshes" and refers to the usual habitat of this species. *Capillaceum* is Latin for "hairy" or "long filaments" and is descriptive of the thread-like leaves. The common name refers to a resemblance, but no relationship, to the true bishop's weed.

Bittercress *Cardamine pensylvanica*
The Greek words which make up the generic name, *kardio* and *damao,* mean "to subdue the heart." It refers to the use of this plant in antiquity in the treatment of heart ailments. The specific name is that of the state from which an early specimen was described.

This is one of the smallest of the spring cresses and is not as bitter as its name suggests. In fact, it is used as a salad plant and has a watercress flavor.

Bitterroot. See Lewisia

Bittersweet *Solanum dulcamara*
Solanum is the old Latin name for this and related plants. It derives from *solamen,* meaning "comforting," and refers to the sedative properties of some species. *Dulcamara* is Latin for "bittersweet." The coral-red berries of this plant have a sweetish taste, then a bitter one. The entire plant is poisonous, however.

Bittersweet, Shrubby *Celastrus scandens*
Celastrus is an old Greek plant name that has been applied to this genus in modern times. *Scandens,* meaning "climbing," aptly describes this vine. A confusion of its seeds with those of an herbaceous bittersweet (*Solanum*) led to the adoption of the same common name for the two unrelated plants. The boiled bark of this vine is sweet and palatable and can be used as an emergency food. The orange and scarlet seeds and capsules are attractive home decorations in the late fall.

Bittercress

Shrubby Bittersweet

Black Alder. See Winterberry

Blackberry *Rubus spp.*
The generic name, from the Latin *ruber* for "red," relates to the color of some berries in this genus, particularly the red raspberries. The common name denotes the color of the ripe berries in one division of this genus.

Gray lists 205 species of *Rubus* in North America. Well-known members of the blackberry genus are dewberry, raspberry, loganberry, cloudberry, salmonberry, and wineberry; all are edible. These are used in pies, jams, jellies, wines, and desserts.

Blackberry-lily; Leopardflower *Belamcanda chinensis*
The name of this genus is the Latinization of the East Indian vernacular name. *Chinensis* reflects its Chinese origin. The seed capsule splits open when ripe to expose a cluster of stuck-together black seeds that resemble a ripe blackberry. The attractive, orange-spotted red flower gave rise to the alternate name, leopardflower.

Black Calla; Solomon's Lily *Arum palaestinum*
Arum is the ancient name of this plant and is believed to be derived from the Biblical Aaron-rod. The specific name alludes to its Palestinian origin. The common name is derived from the Greek, *kalos,* "beautiful." The spathe, or hood, is greenish outside and black purple within. The club-like spadix within is dark colored. The combination of a calla-like flower and black-purple color account for the common name. The name Solomon's Lily arose from the Holy Land origin of this plant.

Black-eyed Susan; Coneflower *Rudbeckia hirta*
This genus honors a father and son team of noted Swedish botanists. Olaf Rudbeck (1660–1740) was a professor of botany at Uppsala University and a teacher of Linnaeus, the father of modern botany.

He founded the botanical gardens at Uppsala and left over 11,000 specimens in his herbarium. He wrote many botanical works, and one book, with the very fanciful title *Elysian Fields,* in which he sought to convince the world that Sweden was the site of Atlantis and the cradle of culture. His son, also Olaf, succeeded his father at Uppsala.

Hirta, Latin for "hairy," is descriptive of the leaves and stem. It is the state flower of Maryland.

The common name arose from the contrast between the dark brown central disk, the "black-eye," and the bright yellow ray flowers. The name cone-flower was suggested by the conical shape of the disk.

Bladder Campion *Silene cucubalus*

Silene is from the Greek *sialon,* meaning "saliva." This refers to a sticky secretion which entraps small flying insects alighting on the plant to which this name originally belonged. *Cucubalus* is an old generic name, the meaning of which is lost.

The common name's propriety is readily apparent to one familiar with the genus. It refers to the much inflated papery calyx that resembles a bladder. The campion part of the name is from the Latin *campus,* "field," which alludes to the common habitat of this white-flowered herb.

Young shoots of bladder campion are good as cooked greens. A puree made from boiled shoots is similar to a spinach puree.

Bladder Campion

Bladdernut *Staphylea trifolia*

Two accounts are extant as to the origin of *Staphylea,* Greek for "cluster of grapes." One is that this name originally was applied to another similar plant. The other is that it referred to the flower cluster of the bladdernut. Each leaf of the bladdernut is divided into three leaflets, hence the specific name.

The inflated bladder containing the nut-like seeds gave rise to the common name. This graceful small shrub, with both attractive flowers and fruit, deserves a place in the garden, especially in shaded locales. The seeds of European species are reported to have been eaten like pistachios; however, we find no record of the food.

Bladderwort

Bladderwort *Utricularia vulgaris*
The Latin word *utriculus,* "small bottle," refers to the insect-trapping bladders found on the stalks of these aquatic plants. The filamentous leaves, submerged in shallow water, are propped up by these tiny bladders. This plant lives in quiet clear waters. It is readily identified by its bladders and small yellow flowers. *Vulgaris* bespeaks its common occurrence. The common name translates simply to "bladderplant."

Blanketflower or Indian Blanket *Gaillardia grandiflora*
The generic name honors a noted French patron of botany, Gaillard de Marentoneau. Either he was well-known on a first name basis, or the one who bestowed the honor felt that his last name posed difficulties as a generic name. *Grandiflora,* Latin for "large flowered," aptly describes the species. The showy flowers suggested an Indian blanket design, hence the two common names for this popular garden flower.

Blazing Star *Liatris borealis*
The derivation of this generic name has been lost in antiquity. *Borealis,* from the Latin, refers to the northern habitat of this species. The common name alludes to the bright blue flower spikes. This denizen of meadows and open woods is worthy of introduction into the flower garden. Many of the 18 species often hybridize naturally, producing much variation in the flowers.

Bleeding Heart, Cultivated *Dicentra spectabilis*
See also Bleeding Heart, Wild. The specific name, *spectablis,* proclaims this plant as "worthy of notice." It produces clusters of bright pink flowers in leafy clumps up to two feet tall.

26

Bleeding Heart, Wild *Dicentra exima*

Dicentra is from the Greek *dis* and *kentros,* meaning "two spurs," and refers to the shape of the flowers. *Eximia* means "exceptional" and alludes to the uncommon flower shape.

The common name originated in Chinese folklore from the fancied resemblance of the red flowers to the heart, with a "drop of blood" between the two tiny spurs. Asiatic specimens were found in China by Robert Frost, a plant collector who in 1846 brought them to the Royal Horticultural Society gardens in London. The American species are found in rich, moist rocky woodlands, mainly in the Appalachian area south of New York.

Wild Bleeding Heart

Bletilla *Bletilla striata*

This orchid genus was named, in diminutive, for Louis Blet, a Spanish apothecary who had a noted botanical garden in Algeciras, Spain, at the end of the eighteenth century. An outdoor orchid preferring partly shady, moist places, Bletilla bears six to ten deep purple flowers. It is hardy as far north as New York.

Bloodroot *Sanguinaria canadensis*

Both the scientific and the common name refer to the red root and the orange-red juice. The Latin *sanguis,* "blood," is the obvious reference. The juice of the root was used by many Eastern Indians for decorative purposes.

The root is poisonous when eaten, but is said to be an aid to digestion if taken as a diluted extract. The specific name *canadensis,* "from Canada," denotes that this plant was originally described from a Canadian specimen.

Bluebeard, Hybrid *Caryopteris incana x mongholica*

The generic name, signifying "nut" and "wing," is descriptive of the seed, which has four winged nutlets. The parents of this hybrid were C. *incana,* meaning "gray," and C. *mongholica,* "from Mongolia." This small shrub bears midsummer spikes of blue flowers, hence the common name.

Virginia Bluebell

Bluebell, Virginia *Mertensia virginica*
This genus honors Frank K. Mertens (1764–1831), professor of botany at Bremen University in Germany. The specific name recalls its original description from a Virginia specimen. This attractive early spring herb with bell-shaped flowers occurs in rich bottomlands.

Blueberry, Highbush *Vaccinium corymbosum*
See also Blueberry, Lowbush.
Corymbosum refers to the flat-topped flower and berry clusters of this swamp-dwelling blueberry. The berries are edible and are used in the same ways as the lowbush species.

Lowbush Blueberry

Blueberry, Lowbush *V. angustifolium*
An ancient name, *Vaccinium* is rooted in the Latin *vaccinus,* "cow," as reflected in the German common name *kuhteke.* The connection between blueberry and cow is obscure. The specific name means "narrow-leaved," and the common name needs no explanation.

Blueberries have long been used in pies, muffins, jams, jellies, and puddings, as well as being eaten as fresh fruit. Wild berries are smaller and tarter than the cultivated ones. This species is found in sandy woods.

Blue Cohosh

Blue Cohosh *Caulophyllum thalictroides*
This lengthy scientific name is from the Greek *kaulon* and *phyllon,* meaning "stem" and "leaf." The stem forms the stalk for both the large divided leaf and the flower stalk. The specific name is Greek for "resembling the meadow rue," which it does.

Cohosh is of Algonquin Indian origin, signifying "it is rough." The deep blue berries on the two- to three-foot stalks are conspicuous in rich woods in late summer and fall.

28

Blue Curls *Trichostema dichotomum*
Of Greek origin, *Trichostema* means "hair-sta-mens" and alludes to the thin filaments of the sta-mens. Each branch divides in two (a dichotomy) in a repeating fashion and ends with a single flower.
These blue flowers are unique, with long, down-ward-curving violet stamens, the origin of "blue-curls." This plant was long considered useful in healing wounds.

Blue-eyed Grass *Sisyrinchium mucronatum*
A highly imaginative botanist named this "pig snout" in Greek. Pigs were observed to eagerly grub for the roots and corms of this herb. The specific name, meaning "sharp-pointed," refers to the pointed petals of this flower.
Blue-eyed grass bears tiny blue violet flowers, amidst linear grass-like leaves; hence its name. It is the smallest member of the iris family.

Blue Curls

Blue-eyed Mary *Collinsia verna*
Early botanists chose this genus to honor Zaccheus Collins (1764–1831), a well-known American natu-ralist and vice-president of the Philadelphia Academy of Natural Sciences. *Verna,* Latin for "spring," de-notes the time of flowering. The pretty blue and white flowers suggested the common name to early settlers.

Blue Flag *Iris versicolor*
According to Greek mythology, Iris was the god-dess of the rainbow and messenger of the gods. *Iris* was chosen for the name of this genus, since it has flowers in almost all hues of the rainbow.
Versicolor, "various colors," refers to the several colors in the blue flag flower. The common name de-scribes the pennant-like appearance of the tall flower spike. The root is a strong emetic and cathartic.

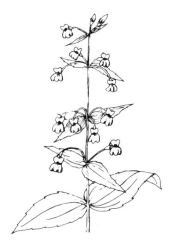

Blue-eyed Mary

Blue Star of Texas *Amsonia tabernaemontana*
Charles T. Amson, an eighteenth-century Glouces-ter, Virginia, physician and plant collector, is given

recognition in this name. A friend of the noted naturalist John Clayton, he collected plant and animal specimens, many of which he gave to Clayton. The species name is the Latinization of the name of Jakob T. von Bergzabern, a sixteenth-century physician to the Count of Palatine at Heidelberg. He was the author of the noted *Neuw Kreuterbuch* (1588). The plant drawings in his book were used later to illustrate Gerard's *Herball*.

The common name is appropriate, since the flowers, common in Texas, resemble a cluster of sky-blue stars.

Bluet

Bluet *Houstonia caerulea*
William Houston (1695–1733) had a short but eventful life. A Scottish botanist, he collected and described plants of Mexico and the West Indies. He first went to the Caribbean area in 1729 as a surgeon with the South Sea Company. As interested in botany as in surgery, he sent seeds and plants to his British associates, especially to the Chelsea Physic Garden. He won recognition as a fellow of the Royal Society. He was commissioned by the Apothecaries Company of London to improve botany and agriculture in Georgia at £200 a year for three years. He died before he reached Georgia.

Caerulea, Latin for "sky blue," is an appropriate specific name for this flower. Bluet, from the French diminutive for blue, is a suitable name for this tiny flower, which often stands only two inches high. Bluets form large colonies beneath mature trees or in other areas where they have no competition.

Boneset *Eupatorium perfoliatum*
This genus commemorates Mithridates Eupator, king of Pontus near the Black Sea, who discovered that one of these species was an antidote for poison. The specific name describes an unusual feature of boneset: the bases of the paired leaves are united around the stem. Long ago this plant had the reputed medicinal virtue of promoting the setting of fractured bones. Boneset flowers in late summer and is commonly found in thickets, low ground, and swampy areas.

Borage *Borago officinalis*
There are two versions as to the origin of the name borage or borago. One is that it is derivative of the Latin *burra,* "rough," a reference to the rough leaves and stems. The other version is that the name sprang from the Arabic words *abu rashsh,* meaning "source of sweat." Either version may be the basis for the Old French *bourrace* and the Latin *borago.*

It was long popular in France as a demulcent to soothe inflammations and as a means to induce heavy sweating, as suggested by the Arabic name. Wine spiked with borage was used as a remedy for melancholy. Borage leaves have been used as a salad green.

Bougainvillea *Bougainvillea spectabilis*
A group of spectacularly showy tropical plants was named in honor of Louis A. de Bougainville (1729–1811), a noted explorer, scientist, lawyer, mathematician, soldier and fellow of the Royal Society of London. To escape his father's profession of notary, he joined the French army and served with Montcalm in Canada. He founded a French colony on Falkland Island in 1764. In 1766 he turned over the colony to Spain. His frigate called at Rio to meet its supply ship. Aboard was a botanist, Commerson, who had collected many plants along the Brazilian coast near Rio de Janeiro. Among these was one purple-flowered vine which Commerson named after his friend Bougainville. Going south through the Straits of Magellan and across the Pacific, Bougainville discovered several islands, one of which was later named for him. He was the first Frenchman to circumnavigate the globe. In 1771 he published a best seller of his day, *Voyage Around the World.*

Spectabilis, Latin for "spectacular" or "showy," is an apt specific name. Bougainvilleas are grown outdoors in California and Florida.

Bouncing Bet; Soapwort *Saponaria officinalis*
The name of this genus is based on the Latin *sapo,* "soap." The bruised leaves have detergent properties and produce a good lather. This quality also is recognized in the specific name, which means "sold by apothecaries."

Bouncing Bet

Bouncing Bet is a short form of the old English bouncing elizabeth. The connection of this name with the flower is lost to us. However, the origin of the second name is clear; soapwort means "soap plant." This white- and pink-flowered plant forms large roadside colonies throughout the eastern United States.

Dwarf Boxwood

Boxwood, Dwarf *Buxus sempervirens*
The name of this low shrub is from ancient Latin and is derived from the Greek name for the box tree. The specific Latin name means "live forever" and alludes to the longevity of the slow-growing species. Most garden examples belong to the variety *nana,* meaning "dwarf."

Brazilian Edelweiss *Rechsteineria leucotricha*
Pfarrer Rechsteiner (1797–1858) was a Swiss-German clergyman who, despite his clerical duties, established a reputation as a botanist, which led to his commemoration in this generic name. The specific name means "white haired" and is descriptive of the overall appearance of the plant. A general resemblance to the alpine edelweiss suggested the common name.

Bridal Wreath *Spiraea vanhouttei*
Spiraea is from the Greek, "something wreathed," and refers to the ancient use of these flowers in garlands, especially on such festive occasions as weddings. L. B. van Houtte (1810–1876) was a noted Belgian nurseryman, horticultural editor, and director of the national school of horticulture. The common name is based on the old Greek name.

Broad Bean *Vicia faba*
See Bean, Yard-long, for the origin of the generic name.
Faba, Latin for "bean," refers to the large, thick beans, often up to four inches long. The common name emphasizes the width of this legume, up to

one inch. Broad beans have been grown in Europe as food for man and beast for many centuries. Their use is gradually spreading in America.

Broccoli, Garden *Brassica oleracea var. botrytis*

Brassica, Latin for cabbage, is also the generic name for many other cabbage-related species, both cultivated and wild. Among these are mustard, cauliflower, and brussels sprouts. *Oleracea* is Latin for "garden herb or potherb." The varietal name means "bunch of grapes" and refers to the compact flower buds which make up this vegetable.

Broccoli is direct from the Italian, meaning "sprout" or "cabbage sprout," and is a diminutive of *brocco,* "splinter." Broccoli is made up of condensed and thickened flower clusters and stems and was developed from an early cabbage type.

Brooklime *Veronica americana*

A pious botanist, we believe, named this flower in honor of Saint Veronica, the woman who is supposed to have wiped the sweat from Christ's face as he carried the cross to Calvary. *Americana* denotes the American origin of this species. The common name is of English origin and was brought here by early colonists. *Lemeke,* basis for the "lime" part of the name, was a denizen of the sides of brook, hence brooklime. This plant is used as a potherb and in salad in spring and early summer and as a preventive of scurvy.

Brook-pimpernel; Water Speedwell *Veronica anagallis-aquatica*

See Brooklime for the derivation of the generic name.

The first part of the specific name is from the Greek, "to delight," and the latter portion is from the Latin, "growing near water." This plant "delights" in growing near water. Its natural habitat includes springs, shallow streams, and wet ditches. We find two conflicting derivations for *pimpernel.* One traces it through the Romance languages to the Latin

Brooklime

33

bipinnula, diminutive for "two winged," which refers to the divided leaves of the original plant so named. The other goes through a similar lingual ancestry to the Latin *piperinus,* "resembling a peppercorn." Some herbalists recommend this as a potherb, but others warn that it is poisonous. It is best left alone.

Scotch Broom

Broom, Scotch *Cytisus scoparius*

The name of this genus is borrowed from an ancient Greek plant name, apparently picked at random. *Scoparius* means "broom-like." Broom is a very old English plant name with Anglo-Saxon and Germanic antecedents. It has been a badge of humility and the emblem of Brittany. It was reputed to cure many diseases. Henry VIII drank distilled water of broomflowers to protect himself against diseases. In later times the tops were used medicinally as a diuretic, and fresh seed pods are reputed to be intoxicating. The roasted seeds are a good coffee substitute. In cookery, the young flower buds and pods can be pickled with vinegar, spices, and salt. Broom has escaped from cultivation in the United States and is common in sandy soils along the East Coast.

Broomrape, Lesser *Orobanche minor*

"Strangler of vetch" is the almost literal translation of the Greek root words of this generic name. *Orobanche* is a parasite on the roots, not only of vetch, but also of tomato, tobacco, hemp, and broom. The *minor* refers to the relatively small size of the plant. The common name recalls the plant's habit, that is, as a parasite on the broom plant's roots. Since this plant chose a parasitic way of life long ago, it has lost its leaves, which today are represented only by minute scales on the flower stalk.

Browallia *Browallia speciosa major*

This genus honors a noted Swedish churchman and botanist, John Browall, bishop of Abo. He was a staunch advocate of the system of classification proposed by Linnaeus. He wrote two noted works, *Harmony of Plant Fruiting with Animal Reproduc-*

tion (1744) and *Examples of Species Transmutation in the Vegetable Kingdom* (1745). *Speciosa* and *major* describe the Browallias as "showy" and "large." This large group of greenhouse and garden herbs, with blue, violet, and white flowers, originated in South America.

Brussels Sprouts *Brassica oleracea gemmifera*
See Broccoli for the derivation of the generic name.

The varietal name means that this plant reproduces by "gemmae" of leaf buds, which when detached from the mother plant are capable of rooting and having a separate existence.

This cabbage-like plant was named for Brussels, Belgium, where it was grown extensively at one time. In early life the young plant resembles its cabbage forbears. Later the stem buds develop into miniature heads, similar to cabbage heads.

Buckwheat, Climbing False *Polygonum scandens*
Polygonum is from the Greek, meaning "many knee joints," and refers to the numerous thickened joints on the stems of these plants. *Scandens* means "climbing," an obvious trait of this vine. The basic word in the common name is from the German *buch waitzen,* which alludes to the resemblance of the triangular seeds to beechnuts. The adjectives inform us that it is not a true buckwheat and that it is a climbing vine.

Buddleia *Buddleia spp.*
This genus of shrubs honors Adam Buddle, a noted English botanist who died in 1715. About 70 species have been described, many in cultivation. The shrubs bear handsome panicles or globular heads of flowers, in violet, lilac, yellow, and white.

Buffalo Bur. See Horse Nettle

Buffalo-weed. See Ragweed, Giant

Buddleia

Bugbane; Black Cohosh *Cimicifuga racemosa*
Both the common and scientific names refer to the reputed insect-repelling properties of this plant. The Latin name means "to drive away bugs," as does the English name. *Racemosa* refers to the loose spikes of its showy white flowers. Cohosh is an Algonquin Indian word, supposedly meaning "it is rough." This name has been applied to several different plants, apparently in error in some instances. The leaves of bugbane are divided into three parts, and each of these is again divided three ways.

Bugleweed *Ajuga reptans*
The scientific name, meaning "not yoked," refers to the unequal and undivided calyx lobes. *Reptans,* Latin for "creeping," describes the growth habit of this species. *Bugle* derives from French *bugle,* late Latin, *bugula,* and Latin *bugillo,* the original name of this plant. Richard Surflet, an early English commentator, wrote of the bugle, "It is put in drinks for wounds, and some do commonly say, that he that hath bugle and sanicle will scarce vouchsafe the chirurgion a bugle."

Bugleweed

The mild white tubers can be sliced and eaten as a relish, or they can be boiled in salt water as a vegetable. See also *Water Horehound.*

Bugloss, Viper's *Echium vulgare*
Echium, an old Greek name meaning "viper," was applied to this plant at a later date. *Vulgaris,* from Latin, refers to its common occurrence. The common name has a complex derivation, from the French *buglosse,* the Latin *buglossa,* and the Greek *bouglossos,* meaning "ox-tongue." It is descriptive of the shape of the flowers.

Bunchberry

Bunchberry *Cornus canadensis*
The Latin name of the cornelian cherry was later applied to this unrelated American genus. There may be a connection: Both species have hard woods, as the basic Latin root, *cornu,* "horn," suggests. This species was first known from a Canadian specimen.

The bright red berry clusters indicate the origin of the common name. They are tart but palatable, especially as a pudding if lemon and sugar are added. In preparing, strain off the seeds after boiling the berries.

Bunchflower *Melanthium virginicum*
Observant botanists noted that the flower parts become blackish after the flowers fade, so they used the Greek words for "black flower," *melas* and *anthos,* in its generic name. The specific name denotes the origin of the first specimen described. This herb bears panicles, or clusters of small flowers, hence the common name.

Bur Cucumber *Sicyos angulatus*
Sicyos is Greek for "cucumber," to which this wild species is related. The specific name refers to the angular or angled fruit of this vine. This climber bears maple-like leaves and small clusters of inedible burred, cucumber-like fruits.

Burdock, Great *Arctium lappa*
The Greek word for *bear,* "arktos," and the Latin for *bur,* "lappa," make up the scientific name of this weed. The application of the Greek part is not clear. The common name tells us that this is a dock which bears burs.

Young burdock shoots are tender, nutritious, and tasty and resemble the salsify, or oyster plant. Gather burdock in the summer before the flower heads are formed. Boil the shoots twice, first with some soda, then with a little salt. The rootstocks also are edible. Remove the bitter rind and cut the core into pieces; then boil or deep fry them. This weed is abundant in waste places, old fields, and pastures.

Burnet

Burnet *Sanguisorba canadensis*
The Latin generic name, from *sanguis,* "blood," and *sorbere,* "to suck," refers to the reputed styptic quality of the juice of this plant. *Canadensis* refers

to the origin of the first specimen described. The common name derives from old French *burnete,* a diminutive of "brown," which refers to the red-brown flowers of the European species. Our native species is edible, though it is not as delicate as its European relative.

Burnet, Great or Salad *Sanguisorba officinalis*
See also *Burnet.*

Officinalis refers to its status as an herb sold in markets or apothecary shops. This was once a very popular salad plant. The young leaves taste like cucumbers.

Burning Bush *Kochia spp.*

This bright and colorful fall garden plant was named in honor of Wilhelm D. J. Koch (1771–1849), a professor of botany at Erlangen University in Germany. Its common name was suggested by the biblical burning bush that was on fire but not consumed. It alludes to the fiery red color of this plant in autumn.

Burning Bush
Kochia spp.

Burning Bush or Gasplant *Dictamnus albus*

Dictamnus is an old plant name that is applied arbitrarily to this genus. It was named for Mt. Dicte in Crete. *Albus* calls our attention to the loose spires of white flowers.

Both common names relate to an unusual characteristic of this plant. On hot summer days the volatile oil vaporizing from the flowers will give a brief flash when lit with a match. This long-lived border plant has lemon-scented foliage.

Burning Bush; Strawberry Bush *Euonymus atropurpureus*

We are uncertain as to the significance of the Greek generic name, meaning "of good repute." Possibly this refers to the medicinal use of the root bark, which at one time was prescribed as a cathartic. The six-syllable species name means "dark- or blackish-purple," and refers to the color of the petals. Both common names reflect the attractive purple capsule and scarlet seeds, often gathered in the fall for home decoration. The seeds are poisonous if eaten.

Bur-reed *Sparganium americanum*
The generic name of this plant is the old Latin word for bur-reed. *Americanum* distinguishes it as a North American species. The common name derives from its round, prickly, bur-like fruits. This aquatic plant bears fall tubers that can be used as a starchy vegetable. They are small, and it takes some time to gather a pound or two.

Bur-reed

Bush Clover *Lespedeza violacea*
This well-known genus was named by the French botanist Michaux in honor of V. M. de Zespedes, the Spanish governor of East Florida (1785–1790). Michaux traveled widely in North America, collecting plants, and was grateful to the governor for his interest and helpfulness. These upright shrubby clover relatives bear attractive violet flowers. Their growth habit suggested the common name.

Butter-and-eggs; Yellow Toadflax *Linaria vulgaris*
The generic name arose from the resemblance of these leaves to those of flax. The Latin name, based on *linum,* "flax," means "like flax." *Vulgaris,* Latin for "common," is appropriate for this widespread roadside denizen. The old common name, butter-and-eggs, denotes that each flower has two distinct shades of yellow that resemble butter and egg yolk. This plant is the ancestor of the cultivated linarias.

Butter-bur. See Sweet Coltsfoot

Buttercup, Aborted *Ranunculus abortivus*
A close examination of the seeds of any buttercup reveals their resemblance to a small frog, the Latin meaning of the generic name. *Abortivus,* meaning

Aborted Buttercup

"imperfect" or "parts missing," is descriptive of this species. Its petals are very small or missing, quite unlike all other buttercups.

Buttercup, Bulbous *R. bulbosus*
See also Buttercup, Aborted.
This buttercup differs from others in bearing a bulbous root. It is quite palatable after boiling. It also loses its bitterness and becomes sweet after it has been dried and stored for about a month.

Buttercup, Common or Tall *R. acris*
See also Buttercup, Aborted.
The specific name, meaning acrid or sharp-flavored, describes the taste of this inedible species. The bright yellow waxy flowers gave the buttercup its common name.

Buttercup, Swamp *R. septentrionalis*
See also Buttercup, Aborted.
The specific name, Latin for "of the north," is descriptive of the northern habitat of this species.

Bulbous Buttercup

Buttercup, White; Water Crowfoot *R. aquatilis*
See also Buttercup, Aborted.
The specific name refers to an aquatic habitat. The alternate name also emphasizes the habitat and the divided leaf, which resembles a crow's foot.

Butterfly-Pea *Clitoria mariana*
The remarkable resemblance of this flower to the clitoris explains the generic name. The pretty blue flowers, almost two inches across, look like the garden sweet pea. The specific name honors the Virgin Mary; the reason for this honor is not known. The common name was bestowed on this pea because of its size, beauty, and resemblance to a butterfly.

Butterfly-Pea

Butterfly Weed *Asclepias tuberosa*
This genus was named for Asklepios, the Greek

god of medicine, in recognition of the importance of some species of this flower in Greek medicine. The Latin word, *tuberosa,* is descriptive of the tuberous roots. This orange-flowered milkweed is particularly attractive to butterflies, hence the common name.

Butterprint. See Velvetleaf

Butterwort *Pinguicula vulgaris*
    The generic name is the Latin diminutive for "fat," the "little fat plant," and refers to the greasiness of the leaves. Gerard, in his *Herball,* states that this plant received its common name because of the "fatness or fullness of the leaves." *Vulgaris,* Latin for "common," attests to the widespread distribution of butterwort. It was long used as a substitute for rennet in curdling milk for cheese-making.

Buttonbush *Cephalanthus occidentalis*
    The Greek name, meaning "head flower" refers to the ball-like head of flowers. *Occidentalis,* or "western," indicates its habitat in North America. The flower head becomes a hard, ball-like seed head. The core was once used as a substitute for buttons, hence the common name.

Buttonbush

Button Snakeroot. See Rattlesnake Master

Cabbage to Cypress Vine

Cabbage *Brassica oleracea var. capitata*
The classical name for cabbage, *Brassica,* is the generic name for over 100 species, including many crops cultivated throughout the world. *Oleracea,* meaning "garden or potherb," attests to the long cultivation of cabbage, for perhaps as long as 2,000 years. The varietal name means "in the form of a head." The common name traces through the middle English *caboche,* the Provencal *caboso,* and the Latin *caput,* or "head."

All the members of this group tend to have thick roots, buds, leaves, and midribs when they are cultivated. Among them are broccoli, cauliflower, kohlrabi, and Brussels sprouts. Several species, including mustard, are the sources of important oils extracted from the seed.

Cactus, Prickly-pear *Opuntia humifusa*
The generic name was originally applied by the Greeks to a plant common around Opuntium in ancient Greece. The specific name, Latin for "trailing," describes this cactus's growth habit. *Cactus,* a word of Greek origin, was used by Theophrastus, a celebrated Greek philosopher.

Our only eastern cactus, the prickly-pear has thick jointed bristly pads, showy yellow waxy flowers, and a red pulpy fruit. The cactus bristles can be painful in one's fingers; use heavy gloves when picking the fruit.

Prickly-pear Cactus

Calamus-root. See Sweet Flag

Calanthe *Calanthe spp.*
Calanthe, signifying "beautiful flower" in Greek,

is a popular hothouse orchid with large rose-colored or white flowers. Four species from India and Malaya are the origin of many horticultural varieties.

Calathea *Calathea insignis*
The botanist who described this plant observed that the South American natives used the leaves in basket-making. The Greek word *calathos,* meaning "basket," is the basis for the name he bestowed on this genus of foliage plants. The specific name, Latin for "remarkable" or "notable," distinguishes this as a species from which over forty cultivated varieties have sprung.

Calendula; Pot Marigold *Calendula officinalis*
The long flowering period of the calendula prompted its generic name, from the Latin *calendae,* "throughout the months." The specific name hearkens back to the time when the dried flowers were dispensed by the apothecary for use as a vulnerary, an anti-emetic, and for the removal of warts.

This calendula was commonly planted in large doorstep pots, hence the name pot marigold. Many varieties have been developed, including doubles and singles, as well as white, yellow, and deep orange flowers.

Calla, Common *Zantedeschia aethiopica*
Francesco Zantedeschi (1773–1846), an Italian botanist and physician, is remembered by this genus. The species name, and an alternate name, lily-of-the-Nile, recall its original habitat in Ethiopia, in the upper Nile region.

Linnaeus coined the common name, calla, from the Greek word, *kallaia,* or "cock's wattles." This tender calla produces white flowers with spathes up to ten inches long.

Calla, Wild. See Water Arum
Calypso. See Fairy Slipper
Camass. See Hyacinth, Wild

Camellia

Camellia *Camellia japonica*
With this genus name, Linnaeus honored the memory of a Moravian Jesuit priest, George J. Kamel (Latinized to Camellus), who was sent to Manila in 1682. In addition to his missionary duties, he established a pharmacy for the distribution of medicines to the poor and studied avidly the plant and animal life of the region. A keen observer and an able artist, he wrote *Plants of Luzon in the Philippines* and illustrated it with 260 drawings. He also sent many plant specimens and seeds to friends in Europe.
Japonica denotes the country of origin of this species.

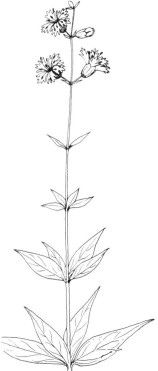

Starry Campion

Campion, Starry *Silene stellata*
See Bladder Campion for the derivation of the generic name.
Stellata is Latin for "starry," a name suggested by the appearance of the white flowers with fringed petals. The common name, campion, is from the Latin *campus* or "field," the usual habitat of this species.

Cancerroot *Orobanche uniflora*
This leafless parasitic plant bears a single white flower atop a sticky pale stalk. Its Greek-derived generic name means "choke-vetch," an allusion to its parasitic habit, particularly upon vetches. *Uniflora* means one flowered, and the common name tells us that at one time the roots were considered useful in the treatment of cancer.

Cancerwort. See Cankerroot

Candytuft *Iberis saxitilis*
This genus was named for Iberia, the Roman name for ancient Spain and the original home of several species. *Saxitilis* also refers to the habitat, "growing among rocks." Candytuft arose from the resemblance of the flower stalk to a stick of candy.

Cankerroot; Cancerwort *Kickxia elatine*

This genus honors Jean Kickx (1775–1831), a Belgian apothecary and author of several botanical works, including *Flora of Brussels*. A son, also Jean and a botanist, wrote six botanical works. *Elatine,* Latin for "taller," distinguishes this from other species, which are creeping annuals. This genus was once regarded as a cancer cure; only the name lives on.

Canna *Canna spp.*

The old Latin name *canna,* meaning "reed," once applied to an entirely different group of plants. This name is now associated with stately ornamental plants with strong foliage and showy red or yellow flowers. Over fifty species from tropical America and Asia are known, and many are now under cultivation.

Close observation of the flowers reveals an unusual adaptation. All but one stamen are large and petal-like, and there is only one pollen-bearing stamen per flower. The true petals are tiny, inconspicuous, and often green. Through evolution, the male generative function has been partly sacrificed to provide allurement to insect visitors.

Cantaloupe *Cucumis melo cantalupensis*

Cucumis is an old Latin name for various melonlike vegetables. *Melo* is an abbreviated Latin word for *melopepo,* an apple-shaped melon, which derived in turn from the Greek word for apple-melon or large melon.

The final pentasyllabe recalls Cantaluppi Castle, near Rome, once the country residence of Popes. Here this melon was first grown from seeds imported from Armenia.

Cape Hyacinth *Galtonia candicans*

This genus honors Sir Francis Galton (1822–1911), world renowned anthropologist and African traveler. Galton's 16 books and more than 200 papers include works in psychology, eugenics, fingerprinting, heredity, weather mapping, genetics, and anthropology.

The specific name, *candicans,* "hoary" or "white-woolly," is descriptive of the species. The common name denotes the original South African habitat.

Cape Marigold. See Daisy, African

Cape Primrose *Streptocarpus rexii*
A botanist who observed the unique manner in which the ripe seed capsule splits into two valves named this genus with two Greek words meaning "twisted fruit." The species name honors George Rex, from whose South African estate came the first plants to bloom in England (1826). The showy white or blue primrose-like flowers gave rise to the common name.

Caraway *Carum carvi*
All three names have a somewhat uncertain, possibly common origin. The Greek name *karon* led to the Latin *carum,* which in turn is related to middle Latin *carvi,* the old Spanish *alcaravea,* and the Arabic *al-karawiya.* This valuable herb's peregrinations thus can be traced through its linguistic history. Caria, in Asia Minor, is believed to be the original habitat of this species.

Caraway has had interesting uses in addition to its role as a condiment. At one time the seeds were believed to be a sure protection against baldness. Caraway seed oil, plus sugar, in alcohol was routinely used to help relieve labor pains. Today caraway is counted among the popular condiment plants, and it has escaped locally (been established as a wild plant) from cultivation.

Cardinal Climber. See Cypress Vine

Cardinal Flower *Lobelia cardinalis*
This tall herb with bright, scarlet flowers honors

the memory of Mathias de l'Obel (Matthew Lobel, 1538–1616), a Flemish botanist and physician to James I of England. Lobel traveled widely in Europe in search of new plants and made several discoveries. He wrote two important botanical treatises.

Cardinalis, Latin for "red-colored," aptly describes these showy blooms. In the wild, the cardinal flower prefers wet places near streams and ponds.

Carnation *Dianthus caryophyllus*

The earliest carnations bore flesh-colored flowers that gave rise to the common name, which is derived from Latin *carnatio,* "flesh." The generic name, meaning *divine flower* in Greek, was bestowed by Theophrastus because of its fragrance and beauty. The Greek species name, meaning pink-leaf, refers to its resemblance to garden pinks. The carnation has been under cultivation for over 2,000 years. So many varieties had been developed by the sixteenth century that the herbalist Gerard complained that "their enumeration would be like counting grains of sand."

The pink carnation, a symbol of mother love, was chosen in 1907 as the emblem of Mother's Day. It is the state flower of Ohio.

Cardinal Flower

Carolina Allspice. See Sweetshrub

Carpetweed *Mollugo verticillata*

Mollugo was originally the Latin name for the stickseed, which derived from *mollis,* meaning "soft." The name was later applied to the carpetweed for reasons that are not clear. The species name, meaning "whorled" in Latin, aptly describes the five to six leaves which make up a whorl. The carpetweed is a prostrate plant, forming a mat or carpet on the ground.

Carrion Flower *Smilax herbacea*

This relative of the greenbrier and the catbrier has a stem which lacks sharp prickles. The Greek *smile,*

Carrion Flower

47

meaning "rasping," refers to the sharp prickles, common to most of the genus, which can rend the clothing and skin of the careless passerby. The species name, meaning "herbaceous," distinguishes this species from others which bear woody canes.

The odor of the tiny flowers is very attractive to carrion flies, which explains the common name. The young sprouts are tender and edible in the spring and can be used in salads.

Carrot, Garden *Daucus carota*

The carrot's full scientific name combines an ancient Greek name, *daucus,* with the old generic name *carota,* which is also the root word for the English common name. The carrot was given considerable attention in Holland, where it gained in size and flavor over its progenitors. It was introduced into England during the reign of Elizabeth I.

Carrot, Wild *Daucus carota*

See also Carrot, Garden.

This common weed of roadsides and fields is also known as Queen Anne's Lace because of the lacy flat clusters of white flowers. One need only pull the root for proof of its kinship with the cultivated carrot.

Castor-oil Plant *Ricinus communis*

The economically important seed of the castor-oil plant resembles a tick, "ricinus" in Latin. *Communis,* Latin for "growing in communities," indicates the gregarious growth habit of this species.

This decorative plant bears bean-like seeds from which castor oil has been pressed since the days of antiquity. A powerful purgative, it was once prescribed for an array of human ills, including those of the stomach, spleen, and uterus, and for infestation of intestinal worms. Infusions of the leaves were used as a remedy for rash, itch, eye inflammations, and scabs. In modern times the oil is used in the production of soap, margarine, and lubricants. The ripe seed capsules "pop" open with a noise, shooting seeds many feet in all directions.

Wild Carrot

Catbrier; Greenbrier *Smilax rotundifolia*
See Carrion Flower for the derivation of the generic name.

The species name, Latin for "round-leaved," describes the heart-shaped leaves of this vine. The common name, greenbrier, is based on the shiny green leaves and briery stems. Catbrier alludes to the powerful recurved prickles, suggestive of a cat's claw.

This and other species of *Smilax,* commonly known as catbrier, bullbrier, sawbrier, and greenbrier, are edible in various ways. A flour can be made from the dried, reddish rootstocks, and a cooling drink or tasty jelly can be prepared from the pulverized, strained sediment. Add hot water and sugar, and a delicious tea results. Cool it, and a jelly results. The fresh roots are white, turning red on exposure to air.

Catbrier

The tender young shoots, which grow up to a length of about five inches, can be cut up for salad. They can also be cooked as a vegetable. Select the more vigorous top shoots which are longer and thicker, rather than the weak side shoots.

Catchfly, Night-flowering *Silene noctiflora*
See Bladder Campion for the derivation of the generic name.

The species name, Latin for "night flower," alludes to the fragrant flowers that are open to evening visitors. The common name ties together sticky stems and flies, giving the plant a "catch-fly" role.

Catchweed. See Bedstraw

Cathedral Bells *Cobaea scandens*
This genus name honors the memory of Father Bernardo Cobo (1572–1659), a Spanish Jesuit missionary and botanist in Mexico and Peru for 45 years. He compiled a ten-volume *Natural History of the New World,* which unfortunately was never published.

Scandens, Latin for "climbing," refers to the vigorous growth of this annual, which can reach 25 feet in height. The bell-shaped flowers suggested the common name.

Catnip

Catnip *Nepeta cataria*

Nepeta is an old Latin name for a group of mints which includes the catnip. *Cataria* means "pertaining to cats," a recognition of the feline fondness for this mint. This strong-scented herb is cultivated to some extent, and it is sometimes an escape (grows wild).

Cat's-ear *Hypochoeris radicata*

The name of this genus is from an old Greek word for succory plant. *Radicata,* Latin for "related to the roots," refers to the thick taproot. The common name stems from the resemblance of the hairy leaves to cat's ears. The flowers resemble dandelions, but the leaves readily distinguish the two species.

Cattail, Broadleaf *Typha latifolia*

Typha is an old Greek name for this plant; *latifolia,* Latin for broad-leaved, distinguishes this species. The common name describes the furry brown seed spikes one sees in late summer or autumn. The flower spike is two-parted. The upper part is a slender "tail" of staminate flowers, a bare thin sliver or spear in early summer. The lower and thicker portion make up the pistillate flowers which become the "cattail."

Cattails have many uses that extend over several months. The young flower spikes are good as a cooked vegetable, and the abundant yellow pollen can be shaken into a bag and used to give flavor and color to pancakes. Dormant sprouts, found at the ends of rootstocks, are good in a salad or cooked. The cut thickened rootstock can be roasted or boiled. Finally, in mid-spring the young shoots can be cut off six inches or so below the surface, and the white part cooked into a superior "asparagus." Be sure to remove the coarse outer leaves.

Cattleya *Cattleya spp.*

This popular genus of orchids was named in honor

of William Cattley, an early English horticulturist and naturalist. Forty species native to tropical America are parents of hundreds of horticultural varieties. These are the showiest of all orchids, of great commercial importance for weddings, holidays, and festive occasions. One or more *Cattleya* species blooms during any month of the year. With a proper selection of species, one can have orchids the year round.

Cauliflower *Brassica oleracea botrytis*

See Cabbage for the derivation of *Brassica*.

The three-word scientific name describes different aspects of this popular vegetable.

Oleracea, Latin for "garden herb or potherb," and *Botrytis,* Greek for "like a cluster of grapes," somewhat describes the cauliflower head.

The common name, the work of a Latinist, means "stem of a plant" and "flower." Cauliflower is an edible head of compressed flower buds on a much thickened stem.

The cauliflower will readily hybridize with cabbages and other related plants to form interesting novelties. Visualize a cabbage-cauliflower mix or kohlrabi-cauliflower combination.

Celandine *Chelidonium majus*

According to folklore, the *celandine* begins flowering with the arrival of swallows and ceases with their departure. This led to the generic name, Greek for "swallow." The species name *majus* means "greater," since this is the larger of several species. *Celandine* is the anglicization of the scientific name.

The bright orange juice at one time was used in removing warts. It likewise was prescribed as a cathartic and diuretic and for promoting perspiration.

Celandine, Lesser *Ranunculus ficaria*

See Buttercup for the derivation of the generic name.

Ficaria, "fig-like," refers to its tuberous roots. The common name suggests a resemblance to the true celandine, but on a smaller scale. This creeping plant

Lesser Celandine

51

bears attractive flowers.

Celeriac *Apium graveolens rapaceum*

Apium, the ancient Greek name for celery, is applied to this vegetable. The species name refers to its heavy, rank odor, and the varietal name indicates a resemblance to the garden rape.

Celeriac, which means "resembling celery," produces an edible root instead of edible stalks. It is used in flavoring soups and stews and can be boiled and eaten with white sauce. Celeriac has been cultivated since the sixteenth century.

Celery *Apium graveolens*

See also Celeriac.

In its early history, celery appears to have been confused with or considered as one with parsley. The English word is from the French *celeri,* the Italian *selieri* or *selini,* then from *selinum.* This derived from the Greek *selinon,* meaning parsley.

The wild celery from which our cultivated form was developed has an acrid, pungent flavor and thin leafstalks. Celery produces one of the smallest vegetable seeds; about 70,000 make an ounce.

Celosia. See Cockscomb

Chamomile; White Mayweed *Matricaria maritima*

See also Chamomile, Wild or German.

This species was originally known for its seashore habitat, hence its specific name. It bears large white flowers in May, thus its common name. Both species resemble *Anthemis cotula* (mayweed), but they are not ill-scented.

Chamomile, Wild or German *Matricaria chamomilla*

The ancient use of this plant in treatment of female disorders accounts for the generic name, from the Latin *matrix,* "womb." Both the species and common name are derived from Greek and meaning "ground apple." The connection is obscure. Introduced from Europe as a garden plant, chamomile has escaped from cultivation and is common in waste places and along roadsides. Its leaves are pineapple-scented. Dried flowerheads are used in medicine and in making chamomile tea.

Chard, Swiss *Beta vulgaris cicla*

Beta is Latin for "beet," evidence that chard is allied botanically to the beet. *Vulgaris,* meaning common, indicates its widespread cultivation and use. *Cicla* is the old Latin name for the mangel-wurzel, a closely related beet grown as cattle fodder.

Chard is from the French *carde* and the Latin *carduus,* the name of an unrelated plant applied later to this vegetable. The horticulturists who developed Swiss chard concentrated on the pulpy and thickened leaves instead of the root.

Charlock. See Mustard, Wild

Checkerberry; Wintergreen *Gaultheria procumbens*

This genus recalls the name of Hugues Gaultier, mid-eighteenth-century counsellor and physician to the king of France. He also pursued an interest in botany and prepared a *Catalogue of Plants of France* in 1760. He later accepted a position as physician to the royal governor of French Canada. While in Canada he made an extensive collection of plants and accompanied the Swedish botanist Count Kalm on exploring trips.

The species name from Latin refers to the creeping habit of growth of this plant. The spicy red berry is eaten by birds and animals, and the young leaves and berries, when chewed, provide a pleasant wintergreen flavor. The mature dried leaves make an excellent tea substitute. Let the tea steep overnight for a strong aroma.

Checkerberry

Cheeses; Common Mallow *Malva neglecta*

Malva, Latin for "mallow," is derived from the Greek *malakos,* meaning "softening." This refers to the emollient quality of the extract made of mallow

Cheeses

juice. *Neglecta,* Latin for "disregarded," suggests that this small member of the mallow group was long overlooked or disregarded.

The common name probably originated with country folk or children, since it is aptly descriptive of the flat, rounded seed capsules which resemble miniature cheeses. This creeping mallow bears tiny white or lavender flowers.

Chenille-plant; Red-hot Cattail *Acalypha hispida*

Hippocrates bestowed this name upon a nettle later assigned to this genus, apparently in an arbitrary manner. *Hispida,* Latin for "bristly," is applicable to this species. The common names were suggested by the bright red spikes of the foxtail-like flowers.

Chervil, Garden *Anthriscus cerefolium*

Anthriscus was the Greek name for a related species that later was applied to the chervil. The specific name is Latin for "cherry-leaved." *Chervil* derives from the Anglo-Saxon *cerfille,* the Latin *caerefolium,* and the Greek *chairephyllon,* which means "pleasing leaf." The name alludes to the fragrance of the foliage, long used as a herb in soups and salads. The leaves are used very much like parsley, which they resemble.

Chervil, Wild *Chaerophyllum procumbens*

See also Chervil, Garden.

The generic name is based on two Greek words, *chaero,* to please, and *phyllum,* meaning leaf, and refers to the pleasantly scented foliage.

The species name, Latin for "prostrate," reflects the wild chervil's habit of growth.

A related species, C. *bulbosum,* the tuberous chervil, bears small carrot-like roots and is used much like carrots.

Chickweed, Common *Stellaria media*

The star-like flowers of this genus suggested the

Latin name, which means "star-like." *Media* refers to its intermediate size among chickweed species.

Chickweed is highly regarded in Europe as a substitute for spinach. It should be mixed with a strongly flavored green since it is bland.

Chickweed, Field *Cerastium arvense*
Linnaeus noted the horn-shaped seed capsule, a unique feature of this plant. The generic name derives from the Greek word *keras*, meaning "horn." The Latin word *arvense*, "of the cultivated field," signifies that this plant long has been regarded as a weed. The common name indicates that the seeds are relished by birds and poultry.

Chicory, Wild *Cichorium intybus*
Both the common and generic names are derived from the Latin *cichoreum* and Greek the *kichora*. The Latin *intibus* means "endive," applied here because the leaves are used in the same manner as endive.

Chicory has long been used in salads and also is cultivated for its root, which when ground and roasted is mixed with coffee to enhance the flavor or used as a coffee-substitute. Its flowers are typically blue, hence the other common name, blue sailors.

Field Chickweed

Chicory, Witloof *Cichorium intybus*
See also Chicory, Wild.
Witloof is Danish for "white leaf," a reference to the food use of its blanched leaves. Witloof is chicory grown in a special manner. Ordinary chicory roots are placed in tanbark in a dark building and covered with two feet of manure. In about two weeks the heads have sprouted and are ready for removal. Use as a delicate salad green.

Chinese Evergreen *Aglaonema modestum*
The Greek-derived generic name, meaning "bright-thread," apparently refers to the spadix of this arum-type plant. *Modestum* alludes to the "modest" ap-

pearance of this variegated-leaved tropical house plant. Its common name denotes the relative longevity of the leaves and its Asiatic origin.

Chinese Lantern plant *Physalis alkekengi*

Physalis is from the Greek word *physa,* meaning "bladder." It describes the attractive papery "bladder" enclosing the seeds. The species name is from the Arabic *al-kakanj,* the name for the ground cherry to which this is closely related. The common name arose because of the resemblance of the bladders to a paper Chinese lantern.

Chinese Orchid, Hardy *Bletilla hyacinthina*

Louis Blet was a Spanish botanist and apothecary who developed a noted botanic garden in Algeciras, Spain, in the late eighteenth Century. The purplish flowers of this orchid have some resemblance to the hyacinth, hence the species name. The common name is descriptive of this outdoor orchid, which is originally from China. It does well in half-shaded, moist places.

Chinquapin, Water. See Lotus, American

Chives *Allium schoenoprasum*

The ancient Latin name for garlic is also the generic name for this plant, which is a close relative. The species name is Greek for "rush-leek" and refers to the narrow leaves arising from the bulb.

The common name came into English from the French *cive,* which in turn is from the Latin word *cepa,* meaning "onion." Chives is a popular seasoning for soups and stews and is used as a border plant in the garden because of its attractive heads of blue flowers.

Christmas Cactus *Schlumbergera bridgesii*

This cactus was named in honor of Charles Schlumberger, a French naturalist who served with the Marine Engineer Corps. He investigated aquatic micro-organisms and determined their classification. This cactus blooms around Christmas, hence its name. The species name honors Thomas Bridges (1807–1865), who collected plants in South America and California.

Christmas Pepper; Coral Gem; Tabasco Pepper
Capsicum anuum conoides
An apt generic name, *Capsicum* comes from the Latin word *capto,* to bite, and refers to the "hot" pungency of these peppers. The red, yellow, or purple peppers are produced at Christmas time. *Anuum* denotes that this is an annual, and *conoides* refers to the cone-shaped fruit of some varieties.
The name *pepper* stems from the Latin, *piper.* Tabasco is an area of Mexico where a variety of these peppers originated. Coral gem is a nurseryman's name for a red pepper.

Christmas Rose *Helleborus niger*
Helleborus is based on two Greek words, *helein,* "to kill," and *bora,* "food." It indicates that this plant is deadly if it is eaten. The bruised leaves can cause a skin inflammation, so this plant should be handled cautiously. The white flowers appear in late fall, and even at Christmas time in sheltered locations, hence the common name.

Chrysanthemum *Chrysanthemum spp.*
This popular flower, symbolic of Japan, gets its name from two Greek words meaning "golden flower," an obvious reference to the large golden flowerhead. This very diversified group has long been under cultivation. Scores, if not hundreds of books have been written about it.

Cicely, Sweet *Osmorhiza claytoni*
The generic name, meaning "odorous root," refers to the aromatic root that is suggestive of anise or

Sweet Cicely

licorice. The species is named in honor of John Clayton, a noted early American plant explorer, who also was honored by the generic name *Claytonia* (see Spring Beauty for details). The common name originally was Sweet Cecilia, a popular name of unknown origin.

This member of the parsley family bears fern-like leaves, an umbel of white flowers, and grows two to three feet tall. The stout, fleshy roots contain considerable anise oil and are used as an anise flavoring.

Cigar-flower *Cuphea ignea*

A small curved protuberance at the base of the flower tube gave rise to the generic name, Greek for "curved." *Ignea,* Latin for "fiery-hued," is an apt description of the scarlet calyx tube.

This bedding plant has petal-less flowers, but the calyx tube is bright scarlet, tipped with black and white. The common name derives from the flower shape.

Cineraria *Senecio cruetus*

Senecio, from the Latin word for "old man," was chosen as a generic name because the downy, hoary crown of the seed resembles a white or bald pate. The specific name, translated "bloodred," was the flower color of the species originally described. The common name is Latin for "pertaining to ashes," for the ash-colored down on the leaves.

The *cineraria,* a native of the Canary Islands, has been highly hybridized for use as a cut flower and in the garden. It bears large, velvety leaves and daisy-like flowers in many bright hues.

Cinquefoil; Five-finger *Potentilla simplex*

The name *Potentilla* was originally given to this group because of its potency as a medicine. Since fevers often were blamed on evil spirits, a medicine that reduced fever was looked on as potent against evil spirits. The name, diminutive for "powerful," arose from this belief. The specific name alludes to the single flower on each stem.

The five-parted leaves give this plant its common names in both French and English. The plant looks like a yellow-flowered strawberry.

Cinquefoil, Rough *P. norvegica*
See also Cinquefoil.

This species was believed to have originated in Norway, hence its specific name.

Cinquefoil, Shrubby *P. fruticosa*
See also Cinquefoil.

Fruticosa, Latin for "shrubby," describes the growth habit of this species.

Cinquefoil, Sulfur *P. sulphurea*
See also Cinquefoil.

This species bears flat clusters of sulfur-yellow flowers, each up to one inch in diameter.

Rough Cinquefoil

Citronella Grass *Cymbopogon nardus*
This source of citronella oil, the old dependable mosquito repellant, received its generic name because of the boat-shaped bracts on its flowers and seeds. *Nardus,* a word that occurs in French, Latin, Greek, Hebrew, and Sanskrit, means spikenard.

Citronella is a diminutive, derived from the Greek *kitron* or *citron,* which in turn comes from *kedros,* the "cedar" or "juniper." All of these words refer to aromatics.

Clearweed *Pilea pumila*
See Aluminum Plant for derivation of this generic name.

Pumila, meaning "dwarf," is descriptive of this low-growing plant. This member of the nettle family is "clear" of, or without, stinging nettles, hence the name. It grows in moist, shady spots, often near the stinging nettle. It is edible as a potherb.

Clematis, Garden *Clematis macropetala*

This climbing vine's generic name is derived from the Greek word *klema,* meaning "vine branch." *Macropetala,* Greek for "large petaled," is descriptive of the large, attractive flowers of this native of China.

This is one of a dozen or more of the cultivated forms of clematis, which have been extensively hybridized and improved. Clematis is available in many different colors and is both a climber and a herbaceous plant suitable for a border.

Purple Clematis

Clematis, Purple *C. verticillaris*

See also Clematis, Garden.

The species name means "whorled," a reference to its leaves.

Clematis has three-part leaves and showy mauve flowers. Many attractive garden varieties stem from this species and from hybrids. There are about 150 species, of which twenty are in North America. The showiness of its flowers is due to the colorful sepals, since petals are missing or vestigial. The Jackman variety, introduced in 1862, is still very popular.

Climbing Fumitory; Allegheny Vine *Adlumia fungosa*

This vine was named in honor of John Adlum, an early nineteenth-century American pioneer in viticulture. *Fungosa,* from Latin, is descriptive of the spongy persistent (remains after flower has gone to seed) corolla. See Fumitory for the derivation of the common name.

Clockvine *Thunbergia alata*

Carl P. Thunberg (1743–1828), a young botanist, was sent by Linnaeus to various parts of the world, especially Asia, to collect plants. Later he became a professor of botany at Uppsala University in Sweden. *Alata,* Latin for "winged," refers to the winged leaf stems. This vine, which twines like a clock hand, has attractive buff flowers with purple throats. It is grown around trellises, verandas, and arbors.

Cloudberry. See Blackberry

Clover, Alsike *Trifolium hybridum*
See also Clover, Red.
The specific name refers to the supposed hybrid origin of this clover. Alsike recalls the town in Sweden where cultivation of this species began. It is naturalized (has become widely established) in the United States.

Clover, Buffalo *T. reflexum*
See also Clover, Red.
The Latin name of this species means "bent" or "curved back" and is descriptive of the flower. This clover of the plains once was favored by buffalo, hence the common name.

Buffalo Clover

Clover, Crimson *T. incarnatum*
See also Clover, Red.
Incarnatum, Latin for "flesh-colored," aptly describes this cultivated clover, brought from Europe and now widely established here. The flowerhead is elongated, not rounded as in other red clovers.

Clover, Hop or Yellow *T. agrarium*
See also Clover, Red.
The species name, Latin for "of open fields," identifies the usual habitat of the hop clover, which bears hop-like heads of yellow flowers. On maturing these become brownish and fold down, taking on the appearance of dried hops.

Clover, Rabbit-foot *T. arvense*
See also Clover, Red.
Arvense is Latin for "of the cultivated field," the preferred habitat of this weedy clover. The fuzzy, gray-pink, oblong flowerheads, soft to the touch, give this clover its common name.

Hop Clover

Red Clover

Clover, Red *Trifolium pratense*
Trifolium, Latin for "three-leaved," describes the key characteristic of the clovers. The species name, meaning "of the meadows," indicates the preferred habitat of the red clover. This clover is distinguished by the red-purple heads and the leaflets marked with pale chevrons.

For more than a millenium, Clover played a role in the pagan religious rites of Greeks, Romans, Celts, Druids, and Germanic peoples. The "club" of playing cards is a clover. The word for clover has a common origin in about a dozen languages; for example, the German *klee,* the Anglo-Saxon *claefer,* the Dutch *klaver,* the Danish *klever,* and the Latin *Clava.* The last refers to "clava trinodis" of Hercules.

The young leaves and flowerheads can be used in salad or cooked as a potherb. Clover tea, a wholesome drink, is made by steeping the dried flowerheads in boiling water. Red Clover is the state flower of Vermont.

Clover, White *T. repens*
See also Clover, Red.
The lowliest of clovers, the Latin *repens* refers to its "creeping" habit of growth. This clover bears white flowers and has triangular marks on each leaflet.

Cocklebur; Clotbur *Xanthium pensylvanicum*
The generic name, Greek for "yellow," refers to the yellow sap of this common weed. The species name is that of the state from which the first specimen was described.

The common name combines the German *klette,* "to stick," with the English *bur,* that is, a bur that sticks to one's clothing or hair. Cocklebur is an apparent corruption of the basic name.

Cockscomb; Celosia *Celosia plumosa*
The ash-like appearance of the flower heads of some species accounts for the generic name, which is based on the Greek word for "burnt." *Plumosa*

refers to the feathery or plumed flower head.

It requires little imagination to relate the common name to the red crested flower head. This garden annual, grown for its attractive flower heads and foliage, is a native of the tropics. About 35 species are known, of which some are cultivated.

Cohosh, Black. See Bugbane

Coleus *Coleus blumei*
The name of this popular garden and houseplant is from the Greek word for "sheath." This relates to a unique feature, the manner in which the coleus stamens are joined together by a sheath at the base of the corolla. The species name recalls Karl L. Blume (1796–1862), a noted Dutch botanist.

About 150 species of these showy-leaved members of the mint family have been described. *C. blumei* from Java is the probable parent of many of our cultivated varieties. The coleus is easily propagated from cuttings. The enterprising gardener can produce fascinating and unexpected leaf variations from plants grown from seed. Allow plants to flower and set seed; sow the seeds indoors in the fall or the following spring.

Coleus

Colicroot; Star Grass *Aletris farinosa*
An imaginative botanist, well versed in Greek mythology, gave this plant the name of Aletris, a female slave who ground meal. This alludes to the powdery appearance of these plants.

The species name, *farinosa,* calls attention to the mealy, flour-like character of the root. The root was used as a remedy for colic early in the nineteenth century.

Collards *Brassica oleracea var. acephala*
The close kinship of this leafy vegetable to the cabbage is proclaimed by the generic name, which is fully described under Cabbage. *Oleracea* is Latin for "garden herb" or "potherb." The varietal name tells

us that this is a headless form of cabbage. Collards is believed to be a corruption of the old English word colewort, the name of such cabbage relatives as kale and rape.

Coltsfoot *Tussilago farfara*

The use of this plant as a popular cough remedy in the nineteenth century and earlier explains the generic name, based on the Greek *tussis*, "cough." The species name comes from *folia farfarae*, the apothecary's name for the medicinal concoction made from this plant.

This widespread weed, a native of Eurasia, resembles the dandelion. Its lobed, somewhat heart-shaped leaves are suggestive of a colt's foot. Coltsfoot candy, made from the boiled, sugared root, is a delightful confection, with or without a cough.

Columbine

Columbine *Aquilegia canadensis*

Lively imaginations played a role in the origin of both the common and scientific names of this flower. A botanist saw the form of an eagle in the petals and bestowed the name *Aquilegia,* from the Latin *aquila,* "eagle." A gardener in ancient Rome saw five doves perched on the rim of a dish, and he called this flower a columbine, from *columba,* Latin for "dove." *Canadensis* inform us that a Canadian specimen was used in the first description of this plant. This wild relative of the garden species occurs widely in moist, rocky woods. Columbine is the state flower of Colorado.

Columnea *Columnea schiedeana*

This genus of tropical American shrubs and climbers, of which a half-dozen species are cultivated, honors Fabius Columna (or Colonna, 1567–1647), an Italian botanical writer. We are indebted to Columna for having coined the word petal, for describing many new plants, and for being first to use copper plates instead of woodcuts in illustrating his works. His books are noted for the beauty of the figures and the accuracy of the descriptions.

C. J. W. Schiede (1798–1836) conducted botanical exploration in Mexico, especially the areas of Vera Cruz, Jalapa, and Orizaba. His doctoral thesis was on spontaneous plant hybrids.

Comfrey, European *Symphyton officinale*
An ancient Greek plant name, *Symphyton* means "grown together," a reference to comfrey's reputed healing properties when applied as a poultice to wounds and bruises. Its healing virtues have been proclaimed by herbalists through the centuries. It was brought to America as a medicinal herb and has been widely naturalized. See Comfrey, Wild for the origin of the common name.

The first leaves to appear in the spring can be used as a potherb. A coffee is made of a mixture of the roasted ground roots of comfrey and chicory.

Comfrey, Wild *Cynoglossum virginianum*
An early botanist who noted the resemblance of the comfrey leaves to a hound's tongue bestowed *Cynoglossum* as the generic name. The species name identifies Virginia as a habitat.

Wild Comfrey

The common name stems from the Middle English *cumfirie*, and from the Old French *fegier*, "to cause coagulation of wounds." These alluded to the supposed healing attributes of the original comfrey, a different kind of plant.

Compass-plant. See Rosinweed

Coneflower, Purple *Echinacea purpurea*
The spiny covering of the involucre or base of this flower and the bristly seed heads that follow are the characteristics on which the generic name is based. These spines suggested the Greek word *echinos,* meaning "hedgehog," to the early botanist who named this plant. *Purpurea,* for "purple," distinguishes this from other coneflowers. The cultivated form often bears flowers six inches in diameter.

Purple Coneflower

65

Coneflower, Yellow *Rudbeckia laciniata*

See Black-eyed Susan for derivation of the generic name.

The species name, a Latin adjective, is descriptive of the leaves, which are cut deeply into narrow lobes. *Coneflower* alludes to the cone-shaped central or disk part of the flower.

Coontail, Common *Ceratophyllum demersum*

The person who named this plant noted that the tiny individual leaves have a horn-like appearance. This led to the generic name which in Greek means "horn-leaf." The species name denotes the fact that these leaves are submerged in water.

This aquatic plant, found in quiet pools and ponds, bears linear leaves in dense masses. From a distance these resemble a raccoon's tail, thus the common name.

Coralbells *Heuchera sanguinea*

This popular garden flower is named for Johann von Heucher (1677–1747), a professor of medicine at Wittenberg University in Germany and later physician-in-ordinary at Dresden. He was author of several books on medical botany and a volume on plant lore entitled *Of Plants in Mythological History*.

Sanguinea, "blood-red" in Latin, aptly describes the red bell-shaped flowers that look like coralbells.

Coralberry *Symphoricarpos orbiculatus*

The bright red berries of this garden shrub are borne in dense clusters, the characteristic which gave rise to the generic name. It is a combination of two Greek words meaning "to bear together" and "fruit." The spherical berries suggested the specific Latin name *orbiculatus*. The common name alludes to the bright coral color of the fruit.

Coral Gem. See Christmas Pepper

Coral-root, Spotted *Corallorhiza maculata*

The full scientific name has been translated from

Coralberry

the Greek into the common name. This parasitic orchid lacks leaves and green pigment. Its large purple flowers bear white lips spotted with red, hence the *maculata,* or "spotted," segment of the name.

Coreopsis; Tickseed *Coreopsis lanceolata*
The generic name, Greek for "bug-like," refers to the bug-shaped seeds of this daisy-like yellow wild-flower. Its lance-shaped leaves account for the specific name. Later naturalists saw the seed as resembling a tick, hence the common English name. Many varieties of coreopsis are in cultivation; all are derived from the wild species.

Coriander *Coriandrum sativum*
The old Latin name of this herb accounts for both its generic and common name. The generic name, from the Greek word for bug, was suggested by the plant's bug-like odor. The species name, meaning "sown as a crop," alludes to its long cultivation. Coriander seeds are used as seasoning and flavoring in pastries, confections, and liquors. It was once highly regarded as a purgative and as an aphrodisiac. This plant is grown in herb gardens, though it is not among the popular herbs, and it is an ingredient of curry power.

Coreopsis

Corn, Sweet; Maize *Zea mays*
Zea, an old Greek name for a grass, was applied as a generic name for sweet corn, which is a member of the grass family. The specific name is from the Spanish *maiz,* which in turn derives from *mahiz,* a Haitian word which Columbus adopted for this grain.

Corn is related to the German *korn,* which may be a corruption of the Latin *granum.* The early English colonists in America learned of its food use from the Indians and named it Indian corn. When the grain was improved during the nineteenth century, the word "Indian" was dropped and the adjectives "sweet" and "field" were substituted. Its only true wild relative is a Mexican grass, *teosinte,* with which it readily hybridizes (a test of genetic kinship).

Corn Cockle *Agrostemma githago*

The Greek words for "field" and "crown" make up the generic name, a reference to the attractiveness of these flowers in a pasture or fallow field and to their use in garlands. *Githago* is an old generic name, the meaning of which has been lost in history. This hairy weed with bright red or purplish-pink flowers was common in corn and grain fields, hence the common name.

Corn Salad; Fetticus *Valerianella olitoria*

The generic name is the Latin diminutive of the verbs "to be strong." It refers to the potent medicinal properties of the original valerian. Corn salad is probably less powerful than its namesake. *Olitoria,* Latin for "of the kitchen," relates to the use of corn salad as a culinary or salad vegetable.

This plant grows freely, almost weed-like, in grain (corn) fields of Europe and is gathered in the spring and early summer for use as a salad and potherb. *Fetticus,* from the Danish "vettik," indicates that it is relished by lambs and other livestock. Corn salad, transplanted to our shores as a stowaway, is a frequent weed on roadsides and in fields.

Gerard, a noted English herbalist, said of this salad plant, "This herbe is colde and somewhat moist, and not unlike in facultie and temperature to the garden lettuce, in steede whereof in winter and in the first months of the springe, it serves for a salade herbe, and is with pleasure eaten with vinegar, salt and oile. . . ."

Corydalis, Garden *Corydalis spp.*

The generic and common names of this plant derive from the Greek word for lark. The early botanists saw a resemblance between the spur of the corydalis and the spur of the lark. These hardy perennials are related to Dutchman's-breeches. Species in cultivation bear yellow, purple, blue, and rose flowers. Corydalis will succeed in half shade, though they prefer full sun.

Corydalis, Yellow *C. flavula*

See *Corydalis, Garden,* for the derivation of the generic name.

Flavula is Latin for "yellowish," the color of the flowers. A number of species of corydalis are under cultivation. These have blue, purple, and rose flowers, as well as yellow.

Yellow Corydalis

Cosmos *Cosmos sulphureus*

As a compliment to this attractive flower, the Greek word for beautiful was bestowed on it. The species name refers to the sulphur-yellow flowers, which are two to three inches in diameter.

About 20 species of cosmos are under cultivation. Most are natives of Mexico and tropical America. *C. bipinnatus,* a popular cultivated species, has white, pink, and crimson flowers and finely divided leaves (i.e., bipinnate).

Costmary *Chrysanthemum balsamita*

Chrysanthemum, Greek for "golden flower," aptly describes costmary. Its specific name means "with a balsam-like odor."

In the Middle Ages, this plant was widely associated with St. Mary, and the names *root of St. Mary* and *costmary* were applied. They are derived from *costum,* the Latin word for "root."

This tansy-scented herb, often used as a potherb or salad plant, is found in herb gardens, and it often escapes from cultivation around cities.

Cotoneaster, Small-leaved *Cotoneaster microphylla*

The generic name is believed to be derived from two words meaning "similar to quince," since the leaves of some species resemble those of quince. The species name means "small-leaved." This low Asiatic shrub, often planted for landscaping, is valued for its scarlet berries and lustrous leaves.

Small-leaved Cotoneaster

Cowbane

Cowherb

Cowbane *Oxypolis rigidior*

This generic name is based on the Greek word for "sharp acid," an allusion to the taste of the juice. The species name is descriptive of the stiff, unbending character of the plant.

Cowbane, meaning "poisonous" or "baneful" to cows, belongs to the parsley family. It prefers swamps, wet woods, and meadows. Its white flowers, deep green leaves, and odor resemble the parsnip. It is a poisonous plant.

Cowherb *Saponaria vaccaria*

See Bouncing Bet for the derivation of the generic name.

Vaccaria is an old Latin plant name, derived from the word for cow. This in turn gave rise to the common name. This annual, naturalized (brought over and now widely established) from Europe, is found in waste places.

Cow Parsnip *Heracleum maximum*

One of the largest members of the parsley family, this genus was named in honor of Hercules, *Herakles* in Greek. This evil-smelling herb grows up to ten feet tall, hence the "maximum" in its name. It is easily identified by its odor, bulging leafstock base, and flower clusters that often are a foot across. Parsnip derives from the Middle English *pasnepe,* the Old French *pasnaie,* the Latin *pastinaca,* and the verb *pastinare,* "to dig up." This refers to its edible roots.

Despite its evil odor, the young leafstalks and stems are as palatable as stewed celery. These are delicious when boiled twice; discard the first water. The large root, when well cooked, tastes like a rutabaga.

Cowslip. See Marigold, Marsh

Cowwheat *Melanpyrum lineare*

This noxious grainfield weed was named "black wheat," as its Greek generic name suggests, because its jet black seeds often were found in threshed grain. The specific name denotes its narrow, linear leaves and the linear petals on its small yellowish-white flowers. The term "cow," like the frequent botanical use of "dog," denotes something false or of less worth than the real thing, in this instance, true wheat.

Cranesbill, Carolina *Geranium carolinianum*

The long, slender beak of the seedpod suggested a crane's bill to early botanists, and so the Greek word for crane, *geranos,* was used as the root of the generic name. By translation it became the common name, too. The specific name refers to the Carolinas, home of the first specimen described. Cranesbill resembles wild geranium, but its flowers are in compact clusters.

Carolina Cranesbill

Crocus *Crocus vernus*

This spring-flowering bulb derived its name from the Greek word for "thread," a reference to the three thread-like stigmas on each flower. *Crocus* is also the Greek word for saffron, the yellow dye derived from these stigmas.

Many of the cultivated forms of crocus stem from *C. vernus,* whose name signifies "spring." Another species, *C. sativus,* is the chief source of saffron. Saffron is used for coloring in confections and cookery and is an essential ingredient in Spanish rice and bouillabaisse. Saffron was once used to dye the robes of royalty. It requires 4,000 stigmas from as many flowers to make an ounce of saffron.

Crocus, Autumn; Meadow Saffron *Colchicum autumnale*

The generic name recalls Colchis, in Asia Minor, where this plant is plentiful. *Autumnale* refers to its fall-blooming habit. Resembling the crocus, *colchicum* bears rose-purple, white, or purple flowers, as well as bicolor or checkered varieties.

It is the source of colchicine, a narcotic and poison. This powerful drug can interfere with plant cell division, causing abnormal multiplication of cells and producing giant plants, flowers, and fruit.

Crossandra *Crossandra undulaefolia*
Two Greek words, descriptive of the fringed anthers, form this generic name. The species name, from Latin, refers to the wavy or undulating leaves. This cultivated evergreen shrub bears spikes of red or yellow flowers.

Crosswort *Crucianella latifolia*
This generic name, diminutive of the Latin word for "cross," refers to its leaf plan; the whorls of four are suggestive of a cross. *Latifolia* describes the relatively broad leaves of this species. The common name derives from the generic name. This rock garden plant originated in the Mediterranean area.

Croton *Codiaeum variegatum var. pictum*
The generic name is the Latinized form of the Malayan vernacular *kodi* or *kodiho*. The specific name alludes to the highly colored, variable ornamental foliage. The varietal name, meaning "painted" or "picture," indicates the great variation in leaf color and form, a factor which accounts for the popularity of the croton as a houseplant. This genus is also the source of croton oil.
Croton in Greek means "tick," which the outline of the seeds resembles.

Crownbeard *Verbesina occidentalis*
This tall sunflower-like perennial carries a generic name which means "resembling a verbena." How it resembles a verbena is obscure. *Occidentalis,* "of the west," refers to the Western hemisphere. The origin of the common name eludes us, though it may refer to the shape of the flower cluster.

Crown Imperial. See Fritillaria

Crown of Thorns *Euphorbia splendens*
Euphorbus, physician to the King of Mauretania

in ancient times, is remembered by this generic name. He appears to have shown some interest in medical botany. *Splendens,* meaning "shining or bright," is descriptive of the bright red flower-bracts.

The common name describes the long stems—flexible and thorny—usable in fashioning a thorny crown. This popular house plant flowers the year-round, but chiefly in winter. The mostly leafless stems grow up to four feet long.

Crown Vetch *Coronilla varia*

The flower umbels of this genus resemble "little crowns," hence the generic name, a Latin diminutive. The species name, meaning "varying," refers to the flower's various shades of pink. The common name indicates a general resemblance to the vetch and alludes to the flower umbels that look like little crowns.

Crown vetch is useful as a hardy herbaceous border plant. Several cultivated species have yellow and purple flowers.

Crown Vetch

Cryptanthus *Cryptanthus acaulis*

These miniature plants produce tiny flowers that nestle in the foliage, hence the Greek-derived generic name meaning "hidden flower." The species name, meaning "stemless," refers to these flowers. The variegated and colorful leaves of this Brazilian air plant are crowded in a rosette. They grow only a few inches high.

Cucumber *Cucumis sativus*

The common and generic names originated from the Greek word for "cucumber," *kykyon,* then through the Latin *cucumis* and the French *cocombre.* Its domestication is revealed in the species name meaning "sown in field or garden." The staminate (male) flowers are large and showy; the pistillate are small.

Cucumber, Wild. See Balsam Apple

Cucumber Root

Cucumber Root; Indian Cucumber *Medeola virginiana*

This genus was named for the sorceress Medea, in recognition of its reputed medicinal virtues. *Virginiana* refers to the state from which the species was described.

The white succulent roots resemble miniature cucumbers, hence the common name. This woodland plant is easily distinguished by the two whorls of four to eight leaves on a slender stalk. The rootstock makes a pleasant nibble and is used in salads and for pickles.

Cudweed, Low *Gnaphalium uliginosum*

The soft, woolly leaves, typical of most of this group, suggest the generic name, which in Greek means "soft down." The specific name, translated "of the marshes," refers to its typical habitat.

The common name arose from the observation that this weed was chewed or cudded by cattle. It was given to cattle that had "lost their cud." Cud is the forage which cattle deposit in the first stomach. Later it is slowly regurgitated and masticated (chewing the cud) for reswallowing into the second stomach.

Culver's Root

Culver's Root *Veronicastrum virginicum*

This generic name is a combination of *Veronica* and *Aster,* since this group has features of both genera. *Virginicum* signifies the state from which an early specimen was described.

This root, long prescribed as a cathartic, is named for a Dr. Culver, who did much to popularize it. This doctor has passed into anonymity, since no records have been found regarding his career. Culver's root is recognized by its spike of small white blooms and leaves in whorls of four to seven.

Cumin *Cuminum cyminum*

The ancient Greek plant name, *kyminon,* is the root of this plant's common, generic, and specific names. An annual herb with finely cut leaves, it belongs to the parsley family. The seeds are used in making curry powder and as a flavoring in pickles and soups. It is easily grown from seeds.

Cupflower *Nierembergia gracilis*
This genus honors a Spanish Jesuit, J. E. Nieremberg (1595–1658), who wrote books on natural history and the marvels of nature. *Gracilis,* Latin for "slender," describes this small, tender annual, grown in rock gardens and in hanging baskets. The white flowers are irregularly cup-shaped, hence the common name.

Cupflower

Cuphea, Clammy; Blue Waxweed; Tarweed *Cuphea petiolata*
See Cigar-flower for the derivation of the generic name.
Petiolata, Latin for "provided with a leaf stalk," distinguishes this from other species. This plant bears clammy, sticky hairs on the stem and on the inflated base of its flower, hence the common names.

Currant, Buffalo *Ribes odoratum*
See also Currant, Red.
This ornamental currant bears yellow fragrant flowers.

Currant, Red *R. sativum*
The Arabic word *ribas,* meaning "a plant or berry with acid juice," is the root of this generic name. *Sativum* is Latin for "cultivated as a crop plant."
Currant is derived from a French word meaning "raisins of Corinth." It was from this port that currants were first introduced into France, and the name is based on the close resemblance of the two dried fruits. The red currant is grown chiefly for its berries, but it also is considered a decorative plant.

Red Currant

Cursed Crowfoot *Ranunculus sceleratus*
See Buttercup for the derivation of the generic name.
Sceleratus, Latin for "polluted" or "defiled," re-

Cursed Crowfoot

fers to the blistering properties of the acrid juice. This quality also accounts for its common name. The leaves are palatable as a potherb if they are boiled twice and the water is discarded each time.

Cyclamen *Cyclamen persicum*

The literature offers two diverse accounts as to the origin of the generic name. One derives it from the Greek word for "circle," an allusion to the spirally twisted flower stalks. The other is that this is the Greek name for sowbread, a bulb which is the favorite food of swine in southern France and Italy. These bulbs added a special flavor to pork products.

Persicum indicates the Persian origin of Cyclamen. There are about a dozen species and scores of horticultural varieties available to the cyclamen fancier.

Gerard, in his *Herball,* writes that the sowbread bulb, "being beaten and made up into little flat cakes, it is reported to be a good amorous medicine to make one in love, if it be inwardly taken."

Cymbidium *Cymbidium eburneum*

An early botanist noted a hollow recess in the lip of this orchid. This led to the generic name, meaning "resembling a boat." The species name, Latin for ivory-white, refers to the color of this epiphytic orchid. Other varieties produce dull purple flowers.

Cypress, Summer. See Kochia

Cypress Vine *Quamoclit vulgaris*

Quamoclit is from the Greek and means "dwarf bean." *Vulgaris* means "common," indicating that it has escaped from cultivation and spread over a broad area. This tropical vine grows to a height of twenty feet and has finely divided leaves that suggest cypress needles. It bears scarlet tubular flowers and is found from Virginia southward.

Cypress Vine; Cardinal Climber *Q. Sloteri*
See also Cypress Vine.

This cultivated hybrid vine was originated by Logan Sloter of Columbus, Ohio, from a cross between *Q. pennata* and *Q. coccinea.* It produces large cardinal-colored flowers.

Dahlia to Dyer's Greenwood

Daffodil. See Narcissus

Dahlia *Dahlia variabilis*
This garden favorite was named in honor of
Andreas Dahl, a pupil of Linnaeus and a Swedish
botanist. He served as a demonstrator in botany at
Abo University in Turku, Finland, and was author
of *Botanical Observations on the Plant System Devised by Linnaeus.*
   There are about ten species of this cultivated perennial, all originating in the Mexican highlands. One
highly variable species, *D. variabilis,* is the parent of
a large number of named varieties. Another popular
species, *D. jaurezii,* parent of the cactus Dahlia, was
named in honor of Mexico's national hero.

Daisy, African; Cape Marigold *Dimorphotheca*
     aurantiaca
Three Greek words meaning "two shapes of fruit"
make up the generic name. This refers to the shape
of the seeds, which vary in form. The Latin species
name, translated "orange-colored," refers to the
flowers. The common names tell us that the original
habitat was South Africa.

Daisy, English *Bellis perennis*
Bellis, signifying "pretty" in Latin, is an apt generic name for this attractive daisy, introduced into
America in colonial times. *Perennis* indicates that it
is a perennial. The large flowers, up to two inches
wide, are white to light pink.

78

Daisy, Painted *Pyrethrum roseum*

The bitter roots of this daisy gave rise to the generic name, based on *pyr,* Greek word for "fire." The specific name refers to the rose-colored flowers. The roots of *P. roseum* were once used as a remedy for fevers. The dried heads of another species, *P. coccineum,* are used commercially as an insecticide.

Daisy, Shasta *Chrysanthemum maximum*

See *Chrysanthemum* for the derivation of the generic name.

The specific name suggests that this is one of the largest species. Actually, the Shasta daisy is a horticultural variety developed in California. It is early flowering and produces an abundance of pure white flowers.

Shasta Daisy

Daisy, Transvaal *Gerbera jamesonii*

The scientific names of this South African daisy-like flower honors two notable botanists. Traugott Gerber, a German naturalist, traveled extensively in Russia in the early eighteenth century and wrote an important monograph on his collection of Russian specimens. Robert Jameson, a Scotsman, went to Durban, South Africa, in 1856. A business man and politician, he served on the Natal Botanic Garden Committee from 1867 until about 1910. In 1885 he and a friend trekked to a new gold-mining operation near Barberton. Unimpressed with the gold prospects of the area, he turned to the plant life of the veldt. He noticed an attractive daisy growing in great profusion. He took some plants with him, which he presented to the curator of the botanical garden. Seeds were later sent to the Kew Gardens in London. The noted botanist J. D. Hooker honored the daisy's discoverer in naming it *Gerbera jamesonii.*

In England this is known as the Barberton daisy, from the locale in South Africa's Transvaal Province where it was found.

Dame's Rocket; Sweet Rocket *Hesperis matronalis*

The marked fragrance of this flower at dusk gave

Dame's Rocket

rise to the generic name, the Greek word for "evening." The same word is the root of the English word, "vespers." The species name, derived from Latin, means "pertaining to the Roman festival of matrons," hence "matronly." The origin of its association with women is not known.

Dame's rocket was introduced into England by Huguenot refugees in the 1500s. It has pink, white, or purple flowers and resembles a phlox, but it has a four-petaled flower, not five, like phlox. It is widely grown in gardens.

Dandelion *Taraxacum officinale*

Taraxacum was the ancient Greek name of a related plant, in turn traceable to an old Arabic and Persian plant name meaning "bitter herb." *Officinale* indicates that dandelion was sold in the marketplace.

The common name, translated "teeth of the lion," refers to the supposed resemblance of the ray flowers to leonine teeth.

Dandelion greens are popular in the spring as a cooked vegetable and as a salad green. These leaves contain a high level of minerals and Vitamin A. Dandelion leaves are also used as an ingredient in soup and can be served creamed or Chinese style. The ground dried roots make a palatable coffee substitute. Dandelion wine, made from the flower heads, is popular with country folks.

A very tasty but little known dish is boiled, properly seasoned dandelion buds. These must be picked in early spring while buds are still esconced in the center of the rosette of leaves. Cut out the entire heart of the rosette for this delicacy.

Dandelion, Dwarf *Krigia virginica*

This denizen of open woods resembles a diminutive dandelion and is often less than four inches high. The genus was named in honor of David Krieg, a German physician who collected plants in Maryland. The type specimen came from Virginia, hence the specific name.

Dangleberry; Tangleberry *Gaylussacia frondosa*
This eastern United States huckleberry was named in honor or Joseph L. Gay-Lussac (1778–1850), a noted French chemist whose work laid the foundations of the food canning industry. He devoted his life to pure and applied science. He invented the hydrometer, alcoholmeter, portable barometer, and steam injection pump. A pioneer balloonist, he ascended to 22,000 feet in 1804.

Frondosa, meaning leafy or like a frond, probably is in reference to the thin, pale leaves. The berries "dangle" in slender-stalked pendulous clusters, hence the common name. Closely allied to the blueberry, this berry is excellent in dessert or for eating out of hand. It is juicy, spicy, and sweet-flavored.

Daphne *Daphne alpina*
Daphne, a river nymph in Greek mythology, was pursued by Apollo. Praying for help, she was transformed into a laurel bush, which became sacred to Apollo. *Alpina* signifies this dwarf shrub as native of the Alps. This rockery (used in rock gardens) plant is popular because of its fragrant white flowers and red berries.

Daphne

Darlingtonia *Darlingtonia californica*
This generic name recalls William D. Darlington of West Chester, Pennsylvania, a noted botanical writer and author of *Florula Cestrica* (1826), a description of the plant life of the West Chester area. The subtitle was: *With Brief Notices on Their Properties and Uses in Medicine, Rural Economy and the Arts."* In 1841 Darlington wrote a *Plea for a National Museum and Botanic Garden at the City of Washington.* Both proposals became reality, though long after Darlington's death.

The specific name indicates California as the habitat of this notable insectivorous plant, introduced to horticulture in 1861. It has five to eight tubular leaves in a rosette. Each leaf has an arched hood with red and green lobed appendages in front, a lure for insects. Once an insect enters the tube in search of nectar, it soon falls to the bottom, unable to reach the entrance because of stiff, downward-pointing hairs blocking the way. The insect falls into a watery grave at the bottom of the tube; its juices are slowly absorbed by the plant.

Dasheen; Taro *Colocasia esculenta*

The generic name is an old Greek plant name, applied much later to this group. *Esculenta* means "edible." *Dasheen* is said to be a corruption of the French *du Chine,* "from China." This starchy tuber is used like a potato in the deep South; the sprouts are used like asparagus in the spring. Dasheen is a close relative of the caladium or elephant's ear.

Dayflower, Asiatic *Commelina communis*

In explaining the origin of this generic name, Linnaeus tells us that, "Commelina has flowers with three petals, two showy blue, and a third inconspicuous, for the two botanists, and the third who died before accomplishing anything in botany." The two noted Dutch botanists were Johann and Kaspar Commelin. Both served as professors of botany in Amsterdam, and each wrote extensively in plant taxonomy (classification) and medical botany. First Johann and then Kaspar served as directors of the Botanical Garden of Amsterdam. The former wrote a catalogue of the plants of Holland; the latter wrote *Flora Malabrica.* Johann was born 38 years before his brother, so Kaspar succeeded Johann in all of the positions held. The third brother died as a young man, without having accomplished anything of note. The dayflower was so named because each flower lasted but one day.

Dayflower, Virginia *C. virginica*

This native dayflower, first described from Virginia, has three blue petals, and the plant is more erect than the preceding. We have seen one report that the fleshy rootstocks are edible when cooked.

Virginia Dayflower

82

Day Lily, Orange *Hemerocallis fulva*

"Beautiful for a day" is the translation of the Greek generic name for the day lily. The attractive flowers open and die in one day. *Fulva,* meaning "tawny orange," is the typical color of this widespread escape (often grows wild) from cultivation.

Each flower stalk bears five or more buds, which open successively for one day. Those that open on a bright clear day may have insect visitors which effect pollination. See following description on the edibility of day lilies.

Day Lily, Yellow *H. flava*

See also Day Lily, Orange.

Flava, meaning "yellow," distinguishes this from the preceding species. It too is a garden favorite which sometimes escapes cultivation.

All daylilies are edible. The peanutlike tubers borne in clusters on the roots can be boiled in salt water or fried. They taste somewhat like sweet corn. The larger buds and newly opened flowers can be mixed in a pancake batter and fried to make a delicious fritter or pancake. The flowers can be cut up, boiled, and added to soups, or can be used as a garnish in meat dishes in the same way that mushrooms are used.

Dead Nettle; Purple Nettle *Lamium purpureum*

See also Henbit.

This member of the mint family owes its generic name to the throat-like appearance of its corolla. The Greek word *lamios,* means "throat," and the Latin *purpureum,* "purple," identifies the red-purplish flowers.

Dead nettle received its name because of its close resemblance to the stinging nettle and the absence of stinging hairs. The purplish hue of the upper leaves assists in attracting insects, an interesting adaptation.

Dead Nettle

Deerberry

Deerberry *Polycodium stamineum*
The Greek generic name means "many little flowers resembling bells." The specific name, "with prominent stamens," is descriptive of the anthers, which extend beyond the tiny greenish-white corolla. According to rural folklore, these berries are relished by deer, hence the common name.

Deergrass. See Meadow Beauty

Delphinium, Garden *Delphinium grandiflorum*
The name of this genus derives from the Greek word for dolphin because of the fancied resemblance of the flower spurs, or partly-opened buds, to a dolphin's head. *Grandiflorum* alludes to the very large flowers of this cultivated species. Among 60 species native to the north temperate zone, this and three others are the parents of most named varieties.

Dendrobium *Dendrobium nobile*
This fanciful Greek name, "tree of life," was given to this genus of orchids in allusion to their aerial habitat, that is, as epiphytes on trees. Over 600 species of these tropical orchids have been described. The species name, Latin for "noticeable" or "excellent," refers to the attractive flowers of this group.

Deptford Pink

Deptford Pink *Dianthus armeria*
See Carnation for the derivation of the generic name.
Armeria is the Latin name for the sea-pinks. The common name refers to the town of Deptford, England. This wild annual is now widely naturalized in America.
Over 250 types of *Dianthus* have been described. The perennials among these are often used in borders and in rock gardens. Although pink and red are the usual colors, white and purple varieties are available.

Desert Candle. See Lily, Foxtail

Deutzia *Deutzia spp.*
This popular ornamental shrub was named in honor of Johann van der Deutz, sheriff of Amsterdam and friend and patron of the noted Swedish botanist Carl Thunberg. About 50 species in the genus come from Asia. Many have been hybridized and introduced into horticulture.

Devil's-bit; Fairy-wand *Chamaelirium luteum*
This Greek-derived generic name meaning "dwarf lily" was suggested by the appearance of this wildflower, which ordinarily grows two to three feet tall, a "dwarf" when compared with some lilies. Another version of this etymology is that "dwarf lily" really refers to the tiny lily-like flowers packed densely on the wand-like spike (an explanation for the other common name). *Luteum,* signifying "yellow," refers to the yellowish-green flower spike of the devil's-bit.

A "bit" or "bite" for the devil is one explanation for the common name. Another explanation, given by Carver in *Travels* (1778), is that it received its name from marks resembling an odd tooth-print that were found on its root. There also is the legend that the short rootstock appears to be bitten off at one end, allegedly the work of a devil intent on destroying its medicinal properties.

Devil's-bit

Devil's Paintbrush; Orange Hawkweed *Hieracium auranticum*
Hieracium, the classical name for another species, was later applied to this genus. The name is based on the Greek word for hawk, ' hieros." Pliny used the juice of the plant in an eye-salve, which improved vision markedly and allegedly gave the user the sharp sight of a hawk. *Auranticum* refers to the deep orange flowers.

The common name, orange hawkweed, is based on the foregoing legend, whereas devil's paintbrush suggests demonic use of this brightly colored flower.

Devil's Walking-stick. See Hercules' Club

Dewberry *Rubus flagellaris*
See Blackberry for the derivation of the generic name.

Flagellaris, Latin for "bearing whip-like stems or shoots," aptly describes the trailing habit of the wild dewberry. The prostrate growth habit of this berry makes it prone to a covering of dew, especially in the morning, hence its common name.

Dewberries are latecomers to cultivation; it was not until the late nineteenth century that they were improved and name varieties offered to the public. Dewberries ripen earlier and bear larger and differently flavored fruit than the blackberry.

Dewberry

Dewdrop *Dalibarda repens*
This genus honors Thomas F. Dalibard (1703–1779) who wrote *Flora of Paris.* He was the first French botanist to plan an important work in accordance with the system devised by Linnaeus. A versatile individual, Dalibard also won recognition as a physicist and as a manufacturer of porcelains. *Repens* refers to the prostrate growth habit of this species.

This low-growing hardy perennial bears strawberry-like flowers and is used in borders and in rock gardens. Dewdrop suggests the diminutive and dew-bedecked character of this species.

Dewdrop

Dill *Anethum graveolens*
Anethum is the Greek word for dill and the root of the word anise. The Latin species name means "heavily scented." Dill derives from an identical German word and in turn has its roots in the Saxon *dillan,* which also means "to lull." A decoction of dill was used to soothe babies to sleep.

A native of Asia Minor, dill is widely cultivated for its aromatic seeds. It is used as a pickling spice for many vegetables, especially cucumbers, beans, and cauliflower.

Dittany *Cunila origanoides*
Cunila, an old Latin name for a mint, was later applied to this genus. The specific name means "resembling origanum," a well-known garden herb. Dittany has a long derivation from the Middle English *dytane,* the French *dictame,* the Latin *dictamnum,* and the Greek *diktamnon.* This name was originally given to a plant abundant on Mt. Dicte in Crete. This mint is easily recognized by the tufts of small purplish flowers and two long protruding stamens.

Dock, Bitter *Rumex obtusifolius*
See also Dock, Curled.
The Latin *obtusifolius* refers to the bluntly or obtusely pointed leaves which distinguish this species. Eaten raw, this dock has a decidedly bitter taste, hence the common name.
The leaves of all of our docks are palatable and nutritious when cooked as greens. They are rich in vitamins A and C. In cooking, discard one or two pots of water to get rid of the bitter flavor. The American Indians prepared a meal from the abundant dock seeds. A good buckwheat-like pancake can be made, but it takes a lot of effort to remove the chaff and grind the seeds. Dock root tea, made from the dried and cut up root, is a gentle laxative and a stimulant to the appetite.

Dock, Curled *R. crispus*
Rumex is the Latin word for dock. *Crispus,* meaning "curly," characterizes the leaves of this species. The name dock was bestowed on this weed because the long thick taproot had the appearance of a dock, or solid portion of an animals tail. As a verb, "dock" is in common usage and refers to the removal of an animal's tail, as with a sheep or dog.

Dodder, Field *Cuscuta pentagona (arvensis)*
Cuscuta, according to one authority, is derived from the Arabic "kushuth." Others state that its origin is unknown, though it was used in the Middle

Curled Dock

87

Ages. *Pentagona* refers to the five-angled seed of this species. A frequently seen older name, *arvensis,* means "of cultivated fields," and refers to its habitat. *Dodder* comes to us from the German word *dotter,* "yolk of an egg," and refers to the yellow color of the flower cluster of some species.

This leafless, parasitic plant bears thread-like twining stems and produces clusters of small flowers. It loses all contact with the ground a short time after germination. Its sucking roots attach firmly to the host plant's stem and absorb nourishment from it. From the distance a heavy infestation appears to be a golden haze.

Field Dodder

Dogbane; Indian Hemp *Apocynum cannabium*

Both the common and generic names signify this plant as poisonous to dogs. The Greek *apocynum* was transferred to this group from an related Old World genus, also poisonous to livestock. *Cannabium,* from Latin, refers to the hemp-like leaves of this species. The dogbane is a widespread weed of fields, shores, and roadsides. It bears small, greenish-white flowers. The root extract was once used as an emetic and cathartic, and Indians used the fibers to make cord, thread, and fishnets.

Dog Fennel *Eupatorium capillifolium*

See Boneset for the derivation of the generic name.

The Latin species name is descriptive of the finely dissected, hair-like leaves. This plant's common name reflects its strong odor and general resemblance to the herb fennel. See also Mayweed.

Dogbane

Dogwood *Cornus florida*

This beautiful ornamental bears a generic name of disputed origin. One authority states that early botanists borrowed the name of the related cornelian cherry. Another states that the name originated from the Latin *cornu,* or "horn," which alludes to the hardness of the wood. There is no dispute regarding the species name which means "flowering." This

Dogwood

88

species bears the most attractive flowers of the entire genus. Here we use "flower" in a very loose sense, since the four "petals" we see are modified leaves or bracts, not part of the flower.

The common name is believed to have originated from dag-wood, or dagger-wood, since this hard wood was used to fashion primitive weapons or skewers. The old English verb *dag* means "to pierce or stab," as with a dagger. Today dogwood is used in making spindles and spools. It is the state flower of Virginia.

Douglasia *Douglasia montana*

David Douglas (1799–1834), a Scottish plant collector and botanist, was sent to North America by the Royal Horticultural Society to collect seeds of plants and trees. He explored the Pacific Coast from California to British Columbia and collected over 300 new kinds of plants.

Montana, "from the mountains," pertains to the Rocky Mountains habitat of this lowly plant. *Douglasia* is used in rock gardens, where it forms compact cushions.

Dracaena *Dracaena marginata*

Dracaena is Greek for "female dragon." The dried juice was supposed to resemble the blood of the female dragon. The red-margined and red-veined leaves gave rise to the specific name. A few species of this tropical plant are in cultivation as indoor ornamentals.

Dragon, Green *Arisaema dracontium*

This generic name is built on two Greek words meaning "of the arum group" and "blood." This signifies a close or "blood" relationship to the arums, which is true of this genus. The "dragon-like" flowers account for the specific and common names.

This relative of the jack-in-the-pulpit bears a greenish-yellow spathe and an elongated spadix. It is found in rich moist woods and along sluggish brooks.

Dragonhead *Dracocephalum moldavica*

The common name is a translation of the Greek

Dragonhead

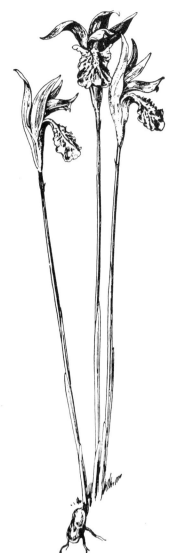

generic name, an allusion to the appearance of the flowers. The original habitat of this garden border plant was Moldavia, a province of Rumania, hence the specific name. The white- and blue-flowered dragonhead belongs to the mint family.

Dragonhead, False *Physostegia virginiana*

The Greek words for "bladder" and "covering" make up this generic name, which refers to the seed capsule covered by an inflated calyx. *Virginiana* identifies the state from which the first specimen was described.

This plant has a superficial resemblance to snapdragon, but its flower structure and square stem identify it as a mint. The common name arose from this similarity. Obedient plant is another common name; when the flowers are turned right or left, they remain in the turned position.

Dragon's-mouth *Arethusa bulbosa*

This beautiful wild orchid was named in honor of the wood nymph Arethusa. The specific name refers to its bulbous root. The basis for the common name is the odd pink flower with three erect sepals and the hood over the blotched and crested lip. It is found sparingly in swamps and spaghnum bogs.

Dragon's-mouth

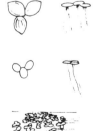

Duckweed *Lemna perpusilla*

This small, free-floating aquatic plant was given the Greek name for waterweeds, *Lemna*. The species name, Latin for "very small," is appropriate for this miniscule plant. The common name derives from the observation that this weed is relished by ducks. It is common on the surface of stagnant and slow-moving waters. One of the smallest of flowering plants, duckweed makes up in numbers what it lacks in size.

Dumb Cane *Dieffenbachia picta*

This genus commemorates the work of J. F. Dief-

Duckweed

fenbach (1794–1847), a botanist and surgeon of Koenigsburg, Germany. Appointed professor of medicine on the Berlin Medical Faculty in 1832, he turned to horticulture late in life and became gardener at the famous Schonbrunn Castle in Austria. The species name refers to the attractive leaves, spotted and feathered with various markings, such as white and yellow blotches.

The acrid juice, when applied to the lips or tongue, causes a temporary loss of speech, hence the common name. This popular house and florist's plant originated in tropical America.

Dusty Miller *Artemisia stelleriana*

The Greek goddess of chastity, Artemis, is recalled by this generic name. There is no apparent connection between this plant and chastity. The species name honors George W. Steller (1709–1746) a German naturalist and explorer in Siberia. A sea cow also is named for him.

The white woolly leaves of this popular garden plant long ago suggested the common name. It frequently escapes from cultivation and is related to the aromatic wormwoods.

Dutchman's-breeches *Dicentra cucullaria*

See Bleeding Heart for the derivation of the generic name.

The species name, meaning "hood-like," refers to the appearance of the flowers when the spurs are uppermost. When spurs are lowermost, the breeches or pantaloons appearance of the flower is very clear.

This delicate spring flower is a worthwhile addition to a wildflower garden. It prospers in a rich moist shaded area.

Dutchman's-breeches

Dutchman's-pipe *Aristolochia durior*

This odd-flowered climbing vine carries an obstetrical generic name, Greek for "best childbirth." A medication made from this plant long ago, was supposed to ease the difficulties of childbirth. The species name, Latin for "harder" or "harsher," is believed to refer to the taste of a decoction prepared

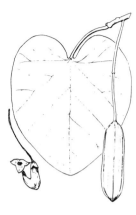

Dutchman's-pipe

from this root. The pipe-shaped flowers gave rise to the popular name of this vine. Birth-wort, another common name, also alludes to its use in childbirth.

Related species are grown as ornamentals because of the odd flowers and the large heart-shaped leaves.

Dyer's Greenwood *Genista tinctoria*

Genista is an ancient plant name associated with this genus, and *tinctoria* is Latin for "relating to dyeing." The flowers of this plant at one time were used to produce a yellow dye, hence the "dyer's" in the common name. The greenwood portion of the name refers to the bright green branches. These take the place of the minute or absent leaves in carrying out the essential function of photosynthesis, the manufacture of food for the plant. This *Genista* is a small, hardy ornamental shrub, belonging to the pea family.

Echeveria to Eyebright

Echeveria *Echeveria glauca*
This genus was named for Atanasia M. Echeverria, a botanical illustrator who accompanied the Sesse-Mocino expedition to Mexico in 1787–1797 and prepared many of the 1,200 drawings of Mexican plants. *Glauca,* Latin for blue or green-gray, refers to the appearance of the rosettes of succulent leaves. *Echeveria* produces small spikes of yellow or pink flowers. It is grown indoors in pots or dishes or bedded out in summer.

Edelweiss *Leontopodium alpinum*
The fancied resemblance of edelweiss flower heads to a lion's foot is the basis of this generic name. The specific name refers to its alpine habitat. The common name, meaning "noble-white" in German, is descriptive of this pure white flower. This native of the high altitudes of Europe and Central Asia is used in America as a rock garden plant.

Eelgrass; Tape Grass *Vallisneria spiralis*
Antonio Vallisnera, for whom this genus was named, served as professor of medicine at the University of Padua, in Italy. He also was a noted naturalist and wrote important works between 1710 and 1730 on the ostrich, chameleon, galls, and the reproductive systems of insects. One work repudiated the biblical theory of the flood. *Spiralis,* indicates the spiraled leaves unique to this species. This submerged aquatic plant bears long narrow leaves, which suggest both common names.

Eggplant *Solanum melongena, var. esculentum*
See Bittersweet for derivation of generic name.

The species name was borrowed from an old name for a kind of melon, which resembles the eggplant in shape. The varietal name, from Latin, translates to "edible."

The common name obviously derives from the egg shape of this vegetable. Botanically, eggplant is related to the potato and tomato. Its ancestral home is believed to be southwest Asia, and its first known culture was in India. Besides the popular purple-skinned variety, there are white, yellowish, striped, and elongated eggplants, the last known as snake eggplant.

Elderberry *Sambucus canadensis*

The name of this genus is traced to the Latin *sambuca,* a Roman harp made of elder wood, which is unrelated to the American berry shrub. The species name records the country from which the first specimen was described. The common name traces to the Old English "ellaern," which was elder.

Elderberries are used to make pies, jam, jelly, wine, rob, chutney, and juice. Elder blow is a delightful dish made of the fresh umbels of flowers. These are dipped in a rich batter and fried in deep fat until light brown. The flowers also can be used in muffins and pancakes. In making pie the berries must first be dried in the sun or in an oven. Elderberires are rich in iron, calcium, and potassium, as well as Vitamin A.

Elderberry

Elecampane *Inula helenium*

Inula is the original Latin name of this flower. The species name, meaning "sun-loving," refers to its preference for sunny locales. The common name also is a Latin derivative and means "field inula."

The cooked roots of elecampane can be candied. John Parkinson, the noted English herbalist, wrote as follows in 1640, "The fresh rootes of Elecampane, preserved with sugar, are very effectuall to warme a cold and windy stomach . . . and it helps the cough, shortness of breath. . . ."

Elecampane

Elephant's Ear *Alocasia spp.*
This generic name is a variant on *Colocasia,* to which it is closely allied. The latter is the Latinization of an Arabic plant name. These tropical herbs from Asia are related to the caladiums. They have very large veined, spotted, and marbled leaves.

Elephant's Ear *Caladium spp.*
Keladi, a local Malayan name for this plant, is the likely origin of the generic name. The common name arose from the resemblance of the huge leaf to an elephant's ear. These tropical plants are grown for their attractive, variegated foliage. Many of our cultivated forms stem from two species, *C. bicolor* and *C. picturatum.*

Elephant's Foot *Testudinaria elephantipes*
The literal translation of this Latin generic name is "resembling a tortoise." The bark of the rootstock cracks into fissures that resemble the pattern of a tortoise shell. The Latin *elephantipes* translates to the common name of this slender twining vine. It is also known as the tortoise plant since the roots are at the surface and their pattern is evident.

Enchanter's Nightshade *Circaea quadrisulcata*
This plant is fabled to have been employed by the enchantress Circe, who was skilled in the use of poisonous plants. The name *Circaea* was applied originally to an Old World species that was poisonous; it later was used as the generic name of nonpoisonous North American plants. The species name is Latin for "four-furrowed" and refers to the plant's stem. This relative of the evening primose received its common name through the influence of its generic name.

Endive *Cichorium endivia*
See Chicory for the derivation of the generic name.
The English name has undergone little change

from the middle Latin specific name, *endivia*. This in turn came from the old Latin *intybum* and from the Greek *entybon*. This leaf salad plant, which originated from chicory, is especially desirable when blanched.

Epidendrum *Epidendrum spp.*

The Greek-derived generic name, meaning "upon trees," alludes to the epiphytic habit of these orchids, that is, growing above ground on trees and vines. Over 500 species have been described in this New World genus. A few are cultivated. The plants draw no nourishment from the supporting tree or vine.

Everlasting. See Strawflower

Everlasting, Garden *Acroclinum helipterum*

The generic name, Greek for "inclined or bent at the top," was bestowed for unknown reasons; however the specific name, Greek for "sun-wing," refers to the tiny plumed (i.e., winged) seed conspicuous in the sunlight. The flowers maintain their color after drying, which accounts for the common name.

Everlasting, Sweet *Gnaphalium obtusifolium*

See Cudweed for the derivation of the generic name.

The Latin specific name, signifying "blunt-pointed leaves," aptly describes them. The white flowers lend themselves to drying and for decorative use, the origin of the common name.

Exacum *Exacum affine*

These herbs once were believed to have the power to expel poisons. From this belief arose the generic name, which is Latin "to drive out." *Affine* means "related or kindred" and refers to other allied species among the thirty or more known from Asia and Africa.

Eyebright *Euphrasia americana*
    The generic name is based upon the Greek word for cheerfulness. This plant had the reputed virtue of clarifying eyesight and curing some eye diseases. Such a cure restored the patient's cheerfulness. The common name is indicative of the American origin of this plant.
    Eyebright is suspected of being partly parasitic on the roots of other plants. The three lower lobes of its white flowers are distinctly notched, a unique characteristic.

Fairy Slipper to Fumitory

Fairy Slipper; Calypso *Calypso bulbosa*
Calypso was a Greek goddess whose name signified concealment. Her name was applied to this orchid because of its concealed habitat. It is difficult to locate specimens in the piney woods where they naturally occur singly or in groups. *Bulbosa* means "bulb-bearing," which is true of this species.

This attractive orchid has a slipper-like lower lip, a yellow crest, and two horns at the toe. The slipper portion is most striking to the observer, hence the common name.

Fairy-wand. See Devil's-bit

Fall Daffodil *Sternbergia lutea*
Count Kaspar M. von Sternberg (1761–1838) was a botanist in Prague, in present-day Czechoslovakia. He also was interested in chemistry, and a crystal silver-iron sulfide was named after him. *Lutea,* a Latin word, refers to the yellow, crocus-like flower of this rock-loving plant.

The fall daffodil, known as the "lily-of-the-fields" in the Bible, grows wild in Israel. In America it is useful as a rockery (rock-garden) plant in hot, dry locales.

Fall Dandelion. See Hawkbit

False Boneset *Kuhnia eupatoroides*
This perennial composite, which resembles the boneset, was named in honor of Adam Kuhn (1741–1817), an American botanist, professor of botany

at the University of Pennsylvania in Philadelphia, and student of Linnaeus at Uppsala University in Sweden, 1761–1765.

The species name means "resembling eupatorium," the genus to which boneset belongs. This perennial bears purplish-white flowers in late summer. It is easily distinguished from boneset by its alternate leaf arrangement.

False Bugbane *Trautvetteria carolinensis*
This tall herb, resembling the bugbane, was named in honor of Ernst R. von Trautvetter, a Russian botanist (1809–1889). A perennial, it is found along stream banks in the southern half of the United States. *Carolinensis* denotes the origin of the first described specimen. Early botanists credited the colony of Carolina as the home of this plant.

False Flax *Camelina microcarpa*
This generic name is from the Greek, meaning "dwarf flax." It was thought to be a degenerate flax, hence a false flax. *Microcarpa,* Greek for "small-seeded," describes the tiny, though numerous seeds this plant produces. This pale yellow-flowered weed is found in grain fields and waste places everywhere in the United States, except the deep South.

False Garlic *Nothoscordum bivalve*
The common name is an exact translation of the Greek generic name. This herb resembles garlic, but lacks its pungent flavor and odor. *Bivalve* has nothing to do with oysters. This Latin name refers to a pair of membraneous bracts beneath the flower head, which look like a pair of tiny bivalve shells.

False Gromwell *Onosmodium virginianum*
The generic name means "resembling onosma," the true European gromwell. *Virginianum* indicates the state from which this species was first described. This member of the forget-me-not family has creamy-

False Flax

white or yellow flowers on a curved flower stalk.

False Heather; Beach Heather *Hudsonia ericoides*
This genus was named in honor of William Hudson (1730–1793), a fellow of the Royal Society and author of *Flora Anglica*. He established the Linnaean system of botany in England and served as director of the Chelsea Botanic Garden. The species name, based on the Greek word for "heath," refers to the heath-like leaves, which are tiny and scale-like and crowded on the low bush.

This genus resembles heather in its foliage and growth habit. Its other common name is derived from its habitat—sand dunes, beaches, and barrens.

False Hellebore *Veratrum viride*
The bestower of this generic name turned to the roots as the unique characteristic. *Veratrum,* Greek for "truly black," refers to the roots of the false hellebore. *Viride,* Latin for "green," is descriptive of the yellowish-green star-shaped flowers. On maturing, these flowers are entirely green.

Hellebore, the classical Greek name of the Christmas rose, was later applied to the true hellebore of Europe, which also bears green flowers.

The false hellebore, a denizen of swamps and wet woods, has heavily ribbed leaves, a distinguishing mark. This plant is violently narcotic and can cause death if taken in quantity. It should not be mistaken for edible marsh herbs.

False Indigo *Baptista australis*
The use of this plant as a substitute for true indigo led to its generic name, which is based on the Greek word for "to dye." *Australis,* Latin for "southern," designates the portion of the United States in which this species occurs. The common name indicates that the dye from some species of *Baptista* is inferior to the true indigo. False indigo has purple-blue flowers and clover-like leaves. Though related to clovers, it is not edible and may be poisonous. See also *Lead Plant.*

False Indigo

False Ipecac; American Ipecac *Gillenia stipulata*

This genus was named in honor of Arnold Gillenius, a seventeenth-century German botanist who had a botanical garden in Cassel. The specific name refers to the pair of leaf-like stipules at the base of each of three-parted leaves. To the uninitiated, each leaf appears to be five-parted, that is, made up of five leaflets. This plant has some of the drug properties of the true ipecac, but it is not related.

False Loosestrife. See Seedbox
False Mallow. See Prickly Mallow

False Nettle *Boehmeria cylindrica*

Another noted German botanist, George R. Boehmer (1723–1803), professor of botany at Wittenburg University, is recalled by this generic name. During his lifetime he published 34 important works. Among these titles were *Handbook of Natural History, Lexicon of the Herb Kingdom, Of Floral Nectar,* and *Of Antisyphilitic Sarsaparillas.* The species name is descriptive of the tubular pistillate calyx (a rather technical point).

The common name originated from this plant's resemblance to the stinging nettle, though it is devoid of stinging hairs.

False Nettle

False Pennyroyal *Isanthus brachiatus*

The botanist who described the first specimen of false pennyroyal noted that the flower parts were regular—alike in size and shape for a mint. This led to the bestowal of a generic name which in Greek meant "regular flowers." The bushy-branched appearance of the herb accounts for the species name, which is Latin for "arm-like," that is, branched at right angles.

False pennyroyal, which does resemble the true pennyroyal, has sticky, entire leaves, and its flowers are on small stalks.

False Pimpernel *Lindernia dubia*

Franz B. von Lindern, who wrote a treatise on

the flora of Alsace, is commemorated in this genus. The species name, Latin for "doubtful," indicates that this species varies from the pattern of characteristics typical of the genus.

Lindernia resembles but is unrelated to the true pimpernel, hence its common name. The early flowers of the false pimpernel are open and insect-pollinated; the late flowers pollinate themselves within the tightly closed flower, a sort of floral incest.

False Solomon's-seal *Smilacina racemosa*

This generic name appears to be something of a misnomer, since it is the diminutive of the Greek word for "scraping" or "thorny." This woodland herb is very smooth, thornless, and bears no resemblance to *Smilax* or greenbrier. The specific name, the Latin *racemosa,* is appropriate, since the flowers occur in racemes or clusters at the tip of the leafy stem. The leaves and habitat are similar to those of the Solomon's seal.

False Solomon's-seal

False Strawberry *Duchesnea indica*

It seems odd that one of the world's greatest experts on strawberry culture should be commemorated with the generic name of the false strawberry. Antoine N. Duchesne, an eighteenth-century horticulturist, discovered many strawberry mutations. Through hybridization, he improved these berries and offered many new varieties to growers. His key work was *Natural History of Strawberries* (1766), followed by the monumental eleven-volume work, *Jardinier Prevoyant* (1770–1781). *Indica* denotes the original habitat, India.

The inedible fruit superficially resembles a strawberry. The yellow flowers readily distinguish this genus from the white-flowered strawberry.

False Strawberry

Fameflower *Talinum teretifolium*

The Greek-derived generic name means "a green branch" and refers to the long-lasting foliage. The species name, Latin for "cylindrical leaves," is a suitable description of the cluster of string-like fleshy leaves at the base of the flower stalk.

One account of the origin of the common name relates to the short-lived attractive pink flowers. The name may have been inspired by the adage, "fame is fleeting."

Fameflower is related to the edible purslane; however, we have found no data regarding its use as a food.

Fameflower

Fanwort, Carolina; Fishgrass *Cabomba caroliniana*

Cabomba is the Latinization of an original Caribbean Indian name. It is widely distributed, occurring through much of south-central United States. The species name denotes that it was first described from a Carolinas specimen.

Cabomba is an aquatic plant with finely divided, thread-like leaves that form a large fan. It is widely used as an oxygenator for home aquaria and is sold in large quantities for this purpose. These facts explain the common names.

Fatshedera *Fatshedera lizea*

This genus received its name in a unique manner. It is a hybrid between the English ivy *Hedera* and *Fatsia,* which is the Latinized version of a vernacular Japanese name for a beautiful foliage plant.

The specific name recalls the feat of the Lize brothers, horticulturists of Nantes, France, who achieved this noted hybridization between two genera in 1912. A fellow horticulturist, André Guillaimin, named the species in their honor.

Fatshedera

Fatsia *Fatsia japonica*

This species has large shining green leaves with lobes spread in finger fashion. It provides a tropical effect, much like the castor-bean plant. A closely allied species, *F. papyrifera,* is the source of Chinese rice paper. *Japonica* refers to the country of origin.

Fatsia

Featherbells *Stenanthium gramineum*

This generic name is traced to two Greek words

meaning "narrow flower," a reference to its small white flowers which have narrow petals. *Gramineum*, a Latin derivation, means "grass-like" leaves.

Featherbells, a member of the lily family, has a single, much-branched flower stalk, bearing tiny flowers, which give it a feathery appearance.

Fennel

Fennel *Foeniculum vulgare* (*or officinale*)

Foeniculum is the diminutive of "foenum," the Latin word for hay. This name was bestowed on fennel because of its hay-like odor. *Vulgare* is Latin for "common" or "ordinary." The other specific name refers to its availability in the marketplace a long time ago. Fennel is cultivated for the aromatic flavoring of its seeds, used as a seasoning and condiment.

A couple of centuries ago fennel was looked upon as an all-purpose medicine. The dried seeds were believed useful in driving worms out of the ears, relieving chest pains, increasing the milk supply of nursing mothers, and in cleansing the liver, gall, and kidneys. Fennel has become widely naturalized in the United States.

Fennel, Florence; Sweet Fennel *F. dulce*

See also Fennel.

This is really a horticultural variety of the common fennel. It has thickened leaf stalks that make a bulb-like enlargement of the stalk near the soil surface. The common names denote the city of its early popularity as well as its sweet flavor. *Fennel* is derived from the original Latin name.

Sweet fennel has a celery-like flavor, but is sweeter. It may be cooked as a vegetable or incorporated in meat dishes, stews, etc.

Fetticus. See Corn Salad

Feverfew *Chrysanthemum parthenium*

See *Chrysanthemum* for the derivation of the generic name.

The species name indicates its origin in Parthenia, an ancient land in western Asia near the Caspian

Sea. The common name is a corruption of two Latin words which mean "to put a fever to flight." A decoction made of feverfew leaves was reputed at one time to have febrifugal properties, that is to put fever to flight. The pinnate or feathery leaves are pungently aromatic. This plant has escaped from cultivation in many localities.

Feverfew

Feverfew, American. See Wild Quinine

Fig, Fiddle-leaf *Ficus lyrata*
See also Fig, Garden.
Lyrata, a Latin word signifying "lyre-shaped," refers to the unusual shape of the large leaves of this ornamental. *Ficus* embraces over 600 species, including such well-known species as the rubber plant and banyan tree.

Fig, Garden *F. carica*
Both the common and generic names derive from the Latin word for fig, *ficus,* which in turn stems from the ancient Hebrew *feg.* This species, parent of many cultivated varieties, is believed to be a native of Caria in Asia Minor. The fig is known to bear in the open as far north as lower Michigan. It is most popular in the southeast and is grown commercially in California.

Figwort *Scrophularia lanceolata*
This plant received its generic name through the influence of the old doctrine of signatures, in vogue in the sixteenth and seventeenth centuries. The roots of some species were thought to resemble scrofulous tumors, so curative properties were ascribed to these roots. The markings were thought to be "signatures" from God. Since this plant was prescribed as a cure for scrofula, it was given an appropriate generic name relating to this fact.
Lanceolata, a Latin word, refers to the lance-shaped leaves of this species. The common name also has a medically related history. An extract of figwort roots was once prescribed as a cure for "figs"

Figwort

105

or "piles," hence figwort or figroot.

Firecracker Plant *Cuphea ignea*
See *Cuphea* for the derivation of the generic name. The species name, signifying "fiery red" in Greek, refers to the bright flowers. The common name also alludes to the bright red flowers between each pair of leaves. This houseplant is a native of Mexico.

Fire-on-the-mountain *Euphorbia heterophylla*
See Crown of Thorns for the derivation of the generic name.

The Greek species name means "various (shaped) leaves" and alludes to the variation in color and shape of the leaves. The common name refers to the bright pink or red leaves which surround each tiny flower cluster. This arrangement resembles that of the poinsettia, which also is a *Euphorbia*.

Fireweed

Fireweed *Epilobium angustifolium*
The bright pink flowers atop the long, immature pods suggested the generic name, which is Greek for "upon the pod." The narrow leaves provided the specific name, meaning "narrow-leaved," in Latin.

This weed is one of the first to spring up in burnt over areas or clearings, hence the common name. Its down-tufted seeds are spread great distances by the wind, thus its primacy in establishing itself in a new habitat. The fireweed grows four to seven feet tall and bears spikes of pink flowers.

The new shoots of this weed make a very palatable substitute for asparagus, and the leafy young stems can be used as a potherb. In England the dried leaves are mixed with India tea to make a refreshing beverage.

Fishgrass. See Fanwort, Carolina

Fittonia *Fittonia verschaffeltii*
Elizabeth and Sarah Fitton, British authors of

Conversations on Botany, are commemorated through the generic name for this popular foliage plant. The specific name honors C. A. Verschaffelt, a nineteenth-century Belgian author of a notable work on camellias.

This low-growing perennial is noted for its brilliant, variegated foliage, especially the red and white venation of its heart-shaped leaves. It does well in deep shade.

Five-finger. See Cinquefoil

Flame Violet *Episcia cupreata*

The strong preference of the flame violet for shady locales led to the generic name, which is Greek for "shady." Similarly, the coppery hue of the leaves naturally led to the appropriate Latin word, *cupreata.* The common name was suggested by the brilliant orange-red flowers, which have some resemblance to violets. This plant is often used in hanging baskets because of its trailing habit and preference for shade.

Flamingoflower *Anthurium andraenum*

The Greek generic name, meaning "tail-flower," refers to the prominent projecting spadix of this flower, which looks like a tail. The honoree for the specific name appears lost to history.

The common name was suggested by the heart-shaped orange-red spathe, which shields the yellowish spadix, the same floral plan as the calla lily or jack-in-the-pulpit.

Flax, Blue *Linum lewisii*

Linum, the Latin name for flax, is derived from the Greek word "linon." This species was named in honor of Meriwether Lewis, the noted leader of the Lewis and Clark Expedition, who discovered this native flax species.

The word *flax* is related to the Dutch "flas" and the German "flacks." This noun originated from the verb "to flay," part of the process of preparation of the fiber.

Flax, Common *L. usitatissimum*
See also Flax, Blue.

This is the flax cultivated for its fiber and occasionally escapes from cultivation. Its Latin specific name means "most common," which may be true in Europe where it has been cultivated the longest. This species is of considerable economic importance, since it is the source of flax, linseed used in animal feed, and linseed oil for paints.

Common Flax

Flax, Garden *L. perenne*
See also Flax, Blue.

This cultivated garden perennial includes sky blue, pink, and white-flowered varieties. The species name indicates that it is a perennial.

Flax, Yellow or Wild *L. virginianum*
See also Flax, Blue.

The type specimen of this flax originated in Virginia. This is one of several yellow-flowered species native to America, found in thickets and clearings.

Fleabane, Daisy *Erigeron annus*
Two Greek words meaning "spring" and "old man" form the root of this generic name. Some of these spring-blooming flowers have a very hoary (old-mannish) appearance at that time. *Annus* indicates that this is an annual plant.

The common name is based upon the belief that this daisy will repel or drive away fleas when the flower heads are dried and placed in a room or near a bed. Its supposed insect-repelling properties remain to be tested in the laboratory.

Daisy Fleabane

Fleabane, Garden *E. aurantiacus*
See also Fleabane, Daisy.

The species name, Latin for "orange-colored," also recalls another name often applied to this flower

—orange daisy. This aster-like annual often blooms through the entire summer. Under optimum conditions, it can produce blooms nine inches in diameter.

Flowering Maple *Abutilon spp.*
This generic name is of Arabic origin, designating a mallow-like plant. The common name is based on its maple-like variegated leaves and its white, pink, or red hibiscus-type flowers. *Abutilon* numbers about 80 species, all native of warm regions. Many cultivated forms are seen in homes, southern gardens, and conservatories.

Fly Poison *Amianthium muscaetoxicum*
Amianthium is a Greek term that calls attention to a glandless perianth, which is unlike several related genera which have glands on the floral envelope. The common name is a translation of the Latin specific name. This plant's juices, especially in the bulb, not only poison flies, but can poison or kill cattle and humans. This member of the lily family has white flowers that resemble the Star of Bethlehem.

Foamflower *Tiarella cordifolia*
Linnaeus, the father of modern botany, noted that the pistil of these flowers was shaped like a small crown. He thus bestowed the Greek word for "little tiara" as a generic name. Taking note of the heart-shaped leaves, he added the Latin *cordifolia* as the species name. The tiny flowers and long slender stamens probably suggested the common name.

Foamflower

Forget-me-not *Myosotis scorpioides*
This universal symbol of loving remembrance bears a generic name of two Greek words meaning "mouseears," which describes the shape of the leaves. The specific name likewise does not evoke pleasant thoughts. This Latin word is translated to "resembling a scorpion" and refers to the appearance of the flower stalk.

Forget-me-not is known by this name in many

Forget-me-not

languages, since it has become an almost universal emblem of friendship, remembrance, and fidelity. Many myths abound as to the origin of this symbolism. The forget-me-not has become naturalized in America and can be found along brooks and in moist places. It is the state flower of Alaska.

Forsythia *Forsythia suspensa*

This popular shrub honors William Forsyth (1737–1804), the Scottish superintendent of the Royal Gardens of Kensington Palace and author of several horticultural treatises. A notable book was his *Treatise on the Cultivation and Management of Fruit Trees.*

Forsyth became a controversial figure because of his claim that "Forsyth's Plaister" was capable of curing all defects in growing trees. This was a mixture of lime, dung, wood ashes, soapsuds, sand, and urine. The British government, in need of sound timber to build its warships, added fuel to the fire by granting him 1,500 pounds to cure defects in oak trees to be used for ship timbers at a later date. Forsyth's prescription for tree ills was exposed as completely valueless, and he finally halted its promotion, whereupon the clamor subsided.

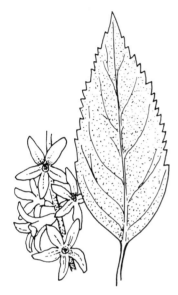

Forsythia

The Latin species name *suspensa* means "hanging" and refers to the dangling golden blooms. Scores of varieties of *Forsythia* are available today for landscape use.

Four-o'clock *Mirabilis nyctaginea*

One can envisage a botanist exclaiming "remarkable" or "wonderful" of a flower which opens late in the afternoon, close to dusk. Such a situation probably occurred, since the Latin generic name signifies the two words just quoted. The species name, Greek for "night-blooming," explains the generic name. The common name provides the approximate hour when the flowers begin to open.

Four-o'clock

Today the gardener can choose from pink, red, purple, yellow, and white varieties of the four-o'clock.

Foxglove, False *Gerardia flava*
See Foxglove, Garden, for the derivation of the common name.

This genus honors one of the greatest figures in the horticultural world, John Gerard (1545–1612). He served as garden superintendent to Lord Burghley, a minister to Queen Elizabeth. He is best known for his *Herball* (1597), one of the earliest botanical works in English. The specific name, *flava,* Latin for "yellow," tells us the color of the flowers.

Foxglove, Garden *Digitalis purpurea*
The finger-shaped corolla is the root of this generic name, derived from "digit," Latin word for finger. *Purpurea,* Latin for "purple," refers to the purple flowers of this species.

The common name has a more involved history. The name originally was fox's glew, an ancient musical instrument garlanded with foxgloves (or other flowers) on festive occasions. The name was gradually altered to fox's glow and eventually corrupted to foxglove. This species is the source of digitalis, a valuable heart stimulant. There are many other cultivated species with yellow, rusty-red, and white flowers.

Freesia *Freesia spp.*
This genus was named in honor of Dr. F. H. T. Freese, a nineteenth-century German botanist of Kiel, and a pupil for Dr. Christian F. Ecklon, who named this genus for Freese.

Freesias are early spring flowering bulbs in most areas, except California. There are many different species that offer great variation in color.

Fringe Tree *Chionanthus virginica*
"Snow-flower" is the translation of the Greek generic name. This alludes to the showy loose panicles of white flowers, which might suggest snow. *Virginica* tells us where the type specimen was found.

The fringed white flowers gave rise to the common

Fringe Tree

name. The fringe tree usually is a shrub rather than a small slender tree. The dried rootbark is used as a gentle laxative and as a diuretic.

Fritillaria; Crown Imperial *Fritillaria imperialis*

It requires a stretch of the imagination to relate a beautiful lily to a dice box. The cup-shaped outline of this lily suggested a dice holder, and so the Latin *fritillaria* was bestowed on this genus.

The species and common name have a similar origin. A large cluster of yellow, orange, or crimson flowers are borne at the top of the two- to three-foot stalk beneath the terminal tuft of leaves. This cluster was thought to resemble an imperial crown.

Frostweed *Helianthemum canadense*

The Greek generic name, meaning "sunflower," is descriptive of the single yellow flower. *Canadense* indicates the locale of the type specimen. The common name arose from the observation that ice crystals form on the leaves during the first frost.

Frostweed

Fuchsia *Fuchsia hybrida*

Leonhard Fuchs (1501–1566), a German physician and botanist, was the author of a botanical work with unusually beautiful wood cuts. He served as professor of medicine at Munich University and Tubingen University in Germany. Two well-known works are the *New Herb Book,* 1543, and *Herbs and Simples,* 1536. These works were considered the most comprehensive of the time on herbs and medicinal plants. He described the newly discovered *Fuchsia,* which was then named in his honor.

Fuchsia hybrida, a popular variety, is believed to be derived from *F. magellanica* and *F. fulgens,* two well-known species. Fuchsias are natives of Mexico, South America, and New Zealand.

Fumitory *Fumaria officinalis*

Both the generic and common names derive from the Latin words for "earth smoke." The ancient

Greeks believed that this plant was produced spontaneously from vapors rising from the earth. The green-gray plant has a smoky appearance from the distance, which aided the earth-smoke theory. Shakespeare spelled the name "fumiter," an in-between form of spelling. *Officinalis* signifies that this plant was sold in apothecary shops.

Fumitory bears spikes of pinkish-purple, crimson-tipped flowers. It is naturalized in America and occurs in waste places and old gardens.

Fumitory

Galinsoga to Groundsel Bush

Galinsoga *Galinsoga parviflora*
This miniature daisy-like flower was named for M. M. Galinsoga, superintendent of the Botanical Gardens of Madrid. *Parviflora,* Latin for "little flower," aptly describes the quarter-inch flower heads, each with five petal-like ray flowers. This Old World species has been widely naturalized in America. The tender upper part of the plant can be used as a potherb in soups and stews. In England the name has been corrupted to Gallant-soldier.

Garden Balm *Melissa officinalis*
See Balsam, Sweet, for the derivation of the generic name.
The species name informs us that at one time it was sold in the market as an edible or medicinal plant. A delightful tea can be made of the dried leaves of this plant. An old recipe calls for the addition of a little lemon juice.

Gardenia *Gardenia jasminoides*
This genus is named for Alexander Garden of South Carolina, a botanist and zoologist who corresponded with Linnaeus. He collected and sent many plants to England, including the magnolia.
During the War of Independence, his Tory sympathies led him to London, where he won recognition in his election as a Fellow of the Royal Society.
The species name signifies "resembling jasmin." A native of China, the gardenia may survive as far north as Washington, D.C. It is widely grown in the South for its large fragrant white flowers.

Gardenia

114

Garlic, Garden *Allium sativum*

Allium is the old Latin name for garlic; *sativum* informs us that it was grown in gardens and fields.

Garlic itself goes back to Middle English "garlek" and the Anglo-Saxon "garlaec." The latter derives from two words, meaning "spear-leak," an allusion to the spear-like leaves. This widely grown herb frequently escapes from cultivation.

Garlic, Meadow or Wild *A. canadense*

See also Garlic, Garden.

Canadense refers to the locale of the first specimen described. Wild garlic is widely distributed throughout North America.

In the Middle Ages and later, it was believed to be a protection against plague and was prescribed as a remedy for many ailments of man and beast. It also was used to protect against the evil eye, demons, and witches. The Romans believed that eating garlic instilled courage on the field of battle. Thus it was planted widely in the wake of Roman conquests.

Garlic insures propagation of the species by several means: aerial bulblets at the end of a long stalk; soft offset bulbs underground; seed produced from a flower stalk; and a hardshell bulb underground.

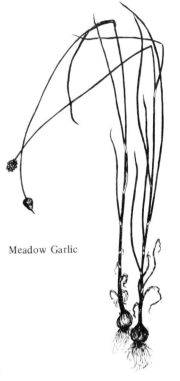

Meadow Garlic

Gasplant. See Burning-bush

Gazania *Gazania splendans*

The generic name of this South African daisy commemorates Theodore of Gaza, in Palestine, a noted Greek scholar who translated the works of Theophrastus from Greek to Latin in the fifteenth century. *Splendans,* Latin for "bright," refers to the showy flowers. This summer bedding plant bears white, yellow, orange, and scarlet flowers with spotted petals. It is a low trailing herb.

Gentian, Bottle or Closed *Gentiana andrewsii*

Gentius was the Illyrian king who discovered the

medicinal virtues of gentian's roots, which were used as a tonic and stomachic, stimulating the appetite and digestion. Gentius was a strong advocate of the use of a decoction of the roots of this plant. The Catawba Indians boiled gentian roots in hot water and applied the extract as a compress for backache.

The species name recalls Henry C. Andrews, an English botanical artist and engraver who published important botanical works early in the nineteenth century.

The corolla of this gentian is almost closed, hence the common name.

The drug gentian is derived principally from the dried root and rootstock of *G. lutea*.

Gentian, Fringed *G. crinita*

See also Gentian, Bottle or Closed.

The specific name, Latin for "hairy," refers to the delicately fringed corolla. This is one of more than 50 species of gentian, many of which are cultivated here and in Europe.

Fringed Gentian

Geranium, Garden *Pelargonium domesticum*

The beak of the geranium seed resembles that of a stork, *pelargos* in Greek, hence the generic name. The Latin species name refers to its cultivated status.

Most geraniums originated in South Africa and were first introduced into England about 1690. One authority lists over 230 species and hundreds of hybrids and varieties. Some are fleshy and tuberous; others have variously scented foliage. Flowers are typically red, though there are purple, white, and yellow varieties.

Early in the nineteenth century a geranium craze developed in England. Strenuous efforts were made to develop showier and fancier *pelargoniums*. This interest was aided and abetted by the landed gentry and wealthy patrons of gardening. A by-product of this interest was the publication of Robert Sweet's *Geraniaceae* in five volumes, with 500 color plates—all on one flower group!

Geranium, Scented *P. capitatum*

See Geranium, Garden, for the derivation of the generic and common names.

The species name, Latin for "growing in a dense head," describes aptly the flower head of this variety of geranium. This popular type has scented foliage as well as attractive flowers.

Geranium, Wild *Geranium maculatum*

See Cranesbill for the derivation of the generic name.

The common name comes from the Greek word for crane, "geranos" because of the resemblance between the seed pod and head and beak of a crane. *Maculatum,* Latin for "spotted," is descriptive of the leaves.

Also known as Herb Robert, it was regarded as a cure for a disease prevalent in Germany, and named for Robert, Duke of Normandy.

Wild geranium was used as an astringent, especially in the treatment of mouth ulcers and sore throat. A decoction was made by boiling the cut up roots of this plant. The dried powdered root was used by northern Indians to promote coagulation of blood in wounds. Its effectiveness was chiefly the result of the tannin content.

Wild Geranium

Gerardia, Purple *Gerardia purpurea*

See the sketch of John Gerard under Foxglove, False.

Purpurea refers to the purple flowers of this species. Some authors refer to this genus as *Agalinis,* a name derived from two Greek words meaning "wonder flax."

Gerardia derives its food by conventional means through leaves and roots, but it is also partly parasitic. Some roots produce suckers at their tips which fasten on to roots of other plants and rob them of nourishment.

Germander; Wood Sage *Teucrium canadense*

This genus was named in honor of Teucer, the

Purple Gerardia

first king of Troy, who used this plant in medicine. *Canadense* refers to the locale of the specimen used in describing the species.

The common name is traced through several languages and alterations of the Greek name for ground oak: the Middle Latin word *germandra,* the Latin-Greek *chamandra,* and the old Greek *chamaidrys.*

In the sixteenth century germander was one of many "strewing herbs," herbs strewn on the floors to provide a refreshing odor to rooms and bedchambers. The other common name, wood sage, arose from its woodland habitat and slight resemblance to sage.

Gilia. See Skyrocket

Gill-over-the-ground

Gill-over-the-ground; **Ground ivy** *Glechoma hederacea*

The gray-green color of the leaves of this lowly herb gives rise to its generic name, the Greek word for "gray-green." *Hederacea,* meaning "resembling ivy," was suggested by its leaf shape and trailing habit.

Gill and Ground ivy are two well-known old English names for this mint. This common herb was known in England by about a dozen names. Gill leaves are almost never more than an inch across, and the plant seldom rises over an inch or two above the ground.

Ground ivy has spread widely over temperate areas of the world. It was well-known as a medicinal herb to the ancient Greeks, and later was highly praised by Gerard. Today this herb has no official standing as a medicinal plant. It is considered a nuisance and a weed when it invades well-kept lawns. Its continued use as a tea is based on its high vitamin C and ascorbic acid content. Gill tea is made from chopped fresh leaves. Pour boiling water over a quarter cup of leaves and let simmer. Sweeten with honey or sugar. The tea is said to be helpful in reducing a cough, in building up a sweat, and in stimulating the appetite.

Ginseng *Panax quinquefolius*

A "panacea for all ills" is the translation of the

Greek words which make up the generic name. This highly exaggerated evaluation was influenced by the high regard for this herb held by the Chinese. Confucius spoke highly of ginseng over 2,000 years ago, and it has been an important part of the Chinese medical armamentarium ever since. *Quinquefolius,* Latin for "five-leaved," describes the five-parted leaf which helps identify this herb. *Ginseng,* from the Chinese *jen-shen,* its vernacular name, is translated freely as "image of man," a reference to the forked root which suggests a human figure.

The supply of wild ginseng in the United States was very much depleted during the nineteenth century because of the constant and heavy demand for the root by Chinese all over the world. In addition, its woodland habitat has been greatly diminished because of lumbering operations. Ginseng is found only occasionally today.

Ginseng

Ginseng, Dwarf. See Groundnut

Gladiolus *Gladiolus spp.*

This popular garden bulb is derived from several of the more than 100 species indigenous to South Africa, where cultivation began about 1807. The name comes from the Latin word meaning "small sword," an allusion to the sword-shaped leaves.

An amateur gardener can expect some pleasant surprises by planting several varieties and allowing insects to hybridize them. Plant the hybrid seeds the following spring, and watch the resultant hybrids when they flower. A gardener interested in more bulbs of the same variety need only set out the cormels formed at the side of the new corm, which develops on top of the old one. These will produce flowers identical with those of the parent bulb.

Glasswort *Salicornia bigelovii*

Two Latin words meaning "salt horn" are the root of this generic name, applied because of the hornlike branches of this genus of salt-marsh plants. The leaves have been reduced to mere fleshy sheathes on the stems and branches. No explanation is forthcoming for this unusual development.

The specific name recalls John (or Jacob) M. Bigelow (1787–1879), a noted figure in American botany. A Boston physician, he gave up his practice for a time in order to serve with Engelmann on the U.S.-Mexican Boundary Commission. He made an extensive botanical collection while engaged in his official duties.

Bigelow returned to Boston where he kept up his botanical interests, collecting extensively while visiting patients in outlying areas. He published a three-volume *American Medical Botany* (1817) and *Florula Bostoniensis* (1814) and left a large herbarium to posterity. In 1820 he chaired the committee which created the American Pharmacopoeia that year.

The common name recalls the early use of this plant in glass-making. It contains a high percentage of alkali, an ingredient in the glass-making process.

Under the colloquial name of chicken claws, the young plant is used as a salty salad ingredient and in pickling. The tender branches are boiled in salt water and then put into a vinegar or spiced oil solution. This plant also was used by country folks in the home manufacture of soap.

Globe Amaranth *Gomphrena globosa*

The name of this genus was borrowed from the Latin name for a type of amaranth. It originated from the Greek word *gomphos,* meaning "club," which alludes to the shape of the flower. The current *Gomphrena* has a globe-shaped flower, as its specific name implies.

Amaranth is the Greek word for "unfading," a reference to the lasting qualities of the flowers. Horticulturists have developed many new varieties, including flowers that are red, white, purple, gold, violet, and striped. The dried heads are widely used in flower arrangements.

Globe Daisy *Globularia bellidifolia*

Globularia, Latin for "little globe," describes the shape of the flower heads. The Latin specific name signifies "daisy-like leaves." The globe daisy is a cushion-like plant, only a few inches high.

Globeflower *Trollius laxus*

This is one of the infrequent generic names that is not Greek or Latin in origin. The derivation is from the German word *trollblume,* which means "globe-like flower." A Latinization provides the name of the genus. *Laxus,* Latin for "loose or open," further describes the flower. The yellow, buttercup-like sepals curve inward to form a small sphere. The tiny petals inside are often absent or go unnoticed.

Globe Thistle *Echinops sphaerocephalus*

The spiny base of the flower head of this thistle suggested the generic name, which in Greek means "like a hedgehog." The species name is Greek for "sphere-headed."

This bold, prickly plant with white woolly foliage grows to a height of eight feet and bears white or bluish heads of thistle-like flowers.

Glory-of-the-snow *Chionodoxa luciliae*

The common name of this early spring flowering bulb is an exact translation of its Greek generic name.

Luciliae was named in honor of Lucile Boissier (1822–1849), who died prematurely while with her husband, the noted Genevese botanist Edmond Boissier, on a collecting expedition in Spain. *Chionodoxa* originated in the mountains of Asia Minor. There are numerous varieties in cultivation, some with blue flowers. *C. grandiflora,* a hybrid, has the largest flowers.

Gloxinia *Sinningia speciosa*

Two notable figures are remembered in this genus. B. P. Gloxin of Strasburg, Germany, was an eighteenth-century botanical writer, later a physician in Colmar, Germany. Wilhelm Sinning was head gardener at the University of Bonn botanical gardens in

Germany. *Speciosa,* meaning "showy," aptly describes this species, the forerunner and leading parent of garden gloxinias.

Because of complex rules of nomenclature, gardeners know this plant as gloxinia, whereas botanists
use the name *Sinningia.*

Goat's-beard *Aruncus dioicus*

An old plant name, its original meaning lost in
antiquity, later was applied to this attractive wild
flower. It has spikes of numerous white flowers and
finely divided leaves.

The specific Greek name, meaning "in two households," indicates that male and female flowers are
borne on separate plants. The white flower stalk is
suggestive of a goat's beard, hence the name. This
perennial herb is sometimes grown as a border plant.

Goat's-beard

Goat's Beard, Yellow *Trapopogon pratensis*

The common name, a translation of the Greek
generic name, was bestowed on this dandelion-like
weed because of the silky plume-like attachment to
the seed, a miniature goat's beard. This performs the
essential function of providing an air-borne lift for
the seed to some distant point. *Pratense,* Latin for
"of the meadow," refers to the usual habitat of this
species.

Goat's Rue *Tephrosia virginiana*

The "ash-colored" leaves of goat's rue inspired the
Greek generic name. The specimen used in describing goat's rue came from Virginia, hence the specific
name.

Rue, a name used to describe several different
plants, traces its origin from the Middle English
rewe, Middle French *rue,* Latin *ruta,* and Greek
rhyte, the last the classical name of a perennial herb.

Fed to nanny goats, this plant was alleged to increase milk production; it was also said to increase
egg production in chickens. This hoary herb bears
yellow and pink pea-like flowers.

Goat's Rue

122

The juice of goat's rue was used by Indians as a fish poison (and it still is so used in Mexico), so there is doubt as to the edibility of any part of this plant, even for goats.

Golden Alexanders *Zizia aurea*
This flower, also known as golden meadow parsnip, was named in honor of a German botanist, Dr. I. B. Ziz. *Aurea,* Latin for "golden," aptly describes the flower umbel. The common name is based upon the Middle Latin name of a related species, *Petroselinum alexandrinum.*

Golden Aster *Chrysopsis mariana*
The bright yellow ray flowers of this aster relative prompted the common name as well as the Greek generic name, which means "golden-appearing." The specific name honors the Virgin Mary, for no special reason other than the namer's piety.

Golden Aster

Golden Club *Orontium aquaticum*
This generic name is of unknown origin; the species name clearly identifies the aquatic habitat of this plant. The spadix, or "club," of this flower bears a mass of minute golden yellow flowers, hence the common name.

The rootstocks are laden with starch in early spring and autumn, and after thorough roasting, they are quite edible, with a potato flavor. The large berry-like seeds, which ripen in midsummer, are edible after boiling.

Golden Coreopsis. See Pot of Gold

Golden Ragwort *Senecio aureus*
See Cineraria for the derivation of the generic name.

Aureus is descriptive of the golden flowers. The common name traces to the Middle English *ragge-*

Golden Club

Golden Ragwort

wort, a reference to the ragged shape of the leaves and perhaps also to the flowers which have only eight or ten ray flowers ("petals") on each flower head.

Golden Ray *Ligularia wilsoniana*
This generic name, Latin for "strap-shaped," describes the ray (or marginal) flowers. The species name honors E. H. Wilson (1876–1930), a Birmingham, England, plant collector and botanist. He traveled extensively in China and brought back about a thousand plants. Later he came to the United States and became director of the noted Arnold Arboretum, near Boston.

Golden ray is the fanciful name given to this garden perennial because of its bright yellow flower, suggestive of a giant, long-stalked dandelion.

Goldenrod, Scented *Solidago odora*
Solidago is based upon the Latin word *solido,* "to make whole," an allusion to the healing qualities once attributed to some species of goldenrod. *Odora,* Latin for "scented," aptly describes this species' anise-like odor.

The common name is clearly descriptive of the flower stalk as a "golden rod." The dried young leaves and flowers make an agreeable tea, one that was very popular with Hessian mercenaries during the American Revolution. For a time during the nineteenth century, this tea was exported to China, where it commanded a high price.

There are over 60 species of goldenrod in the United States. It is the state flower of Alabama, Kentucky, and Nebraska.

Golden Saxifrage

Golden Saxifrage; Watermat *Chrysosplenium americanum*
Long ago this plant was prescribed by herb doctors for ailments of the spleen. An imaginative botanist, in choosing a generic name, combined the Greek words for "gold" and "spleen" to achieve his goal. The species name identifies its American origin.

The common name combines the family identity,

saxifrage, with the color of the flowers. This is a low marsh plant, as suggested by its alternate common name.

Goldenseal; Orange-root *Hydrastis canadensis*

The generic name, based on the Greek *hydro,* meaning "water," alludes to the moist woodland habitat of this group. *Canadensis* refers to the locale of the first specimen described. The thick yellow roots were used in medicine by the Indians and later by frontier settlers. It also can be used to extract a yellow dye. Intensive collecting of roots has almost exterminated this species in many localities where it was once abundant. The common names stem from the use of this species as a source of dye.

Goldenseal was used as a laxative, tonic, stimulant, astringent, and wash for skin eruptions. The mashed root, mixed with fat and applied to the body, served as an insect repellant. This plant was in the U.S. Pharmacopoeia until 1936.

Goldenseal

Goldmoss; Stonecrop *Sedum acre*

This genus took its name from the Latin word *sedo,* "to sit," an allusion to the manner in which this plant "sits" on rocks and walls. The Latin species name, related to our "acrid," describes the sharp or pungent taste of the leaves.

The common name, goldmoss, alludes to the appearance of a mass of the flowers. Stonecrop suggests the manner in which it covers a stony or rocky surface. Goldmoss is widely naturalized in the United States and can become a troublesome garden pest.

Goldmoss

Gooseberry, Garden *Ribes uva-crispa*

Ribes is derived from the Arabic name of a shrub with acidic berries. The species name is Latin for "trembling or quivering berries."

The word *gooseberry* has no known connection with geese. It appears to be a corruption of the French equivalent, *groseille* or of the German *krausbeere.*

"To gooseberry" is an interesting verb. It is the

Garden Gooseberry

125

permission given by parents to a young couple to be together in outdoor privacy for the alleged purpose of picking berries.

Goosefoot, Maple-leaved *Chenopodium hybridum*
The common name is a translation of the Greek-derived generic name. In the case of this species, the leaves look more like maple leaves than like a goose's foot. The species name suggests that this is of mixed or hybrid origin.

This weedy plant bears spikes of tiny green flowers. It prefers clearings and edges of woods.

Goose Grass; Knotgrass *Polygonum aviculare*
See False Buckwheat for the derivation of the generic name.

The specific name, Latin for "relating to small birds," refers to the fact that its seeds are sought after by small birds.

Goose grass may have originated from the disposition of geese to feed upon these seeds. The alternate name, knotgrass, refers to the knotty joints of the stems.

Gourd *Cucurbita pepo ovifera*

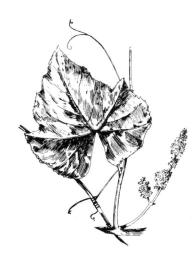

These vines, widely grown for their colorful ornamental gourds, are closely allied to squashes and pumpkins. The generic name is the classical Latin name for gourd; *pepo* is from the Greek *pepon,* the name for pumpkin and related cucurbits. *Ovifera* is Latin for "shaped like an egg," which is true of some gourds.

The common name is traced to the Old French *gourde,* and possibly, in a very circuitous way, to the word *cucurbita.*

Grape, Fox *Vitis labrusca*
Vitis is Latin for grape; *labrusca* is the Latin name for a wild, claret grapevine. The common name has an obvious kinship with the French *grappe,* "bunch of grapes," and with *graper,* "to gather grapes" or

Fox Grape

"Pull off grape bunches with a hook." The latter appears to derive from the German *krampfo,* "a hook."

This is the best and largest wild grape, useful in making jellies, preserves, and wine. The dozen or more species of wild grapes are similarly useful.

Grape Hyacinth *Muscari botryoides*

The sweet scent of some species in this genus suggested the generic name, from the Greek *moschus,* or "musk." The species name, also Greek-derived, signifies "resembling a bunch of grapes," an apt name for this herb. The cluster of ball-like, small purple flowers bears some resemblance to grapes, and the scent might be compared with that of the hyacinth. Together, they make up the common name.

Grape Hyacinth

Grape Ivy *Cissus rhombifolia*

The generic name is the Greek word for ivy, which is applied to this ivy-like climbing vine. The diamond-shaped leaves account for the Latin species name. Those who gave this vine its common name saw the leaves as grape-shaped, hence the name grape ivy.

Grass-of-Parnassus *Parnassia glauca*

Both the generic and common name recall Mount Parnassus in Greece. This delicate, white-flowered member of the saxifrage family is fabled to have originated on the slopes of this mountain. *Glauca,* a Greek word, refers to the whitish bloom on the leaves and stem.

Grass Pink *Calopogon pulchellus*

"Beautiful beard" is the translation of this Greek generic name. It refers to the beard-like fringe on the flower of this attractive native orchid. *Pulchellus,* Latin for "pretty," is the specific name of this denizen of swamps and bogs. Each plant bears a loose cluster of two to ten pink flowers, each with a yellow-crested lip.

Grass-of-Parnassus

Greenbrier. See Catbrier

Gromwell, Corn *Lithospermum arvense*

This weedy flower received its Greek generic name from its "stone-like seeds." *Arvense,* Latin for "of the cultivated fields," refers to its common occurrence as a weed in cultivated fields. Gromwell is derived from the Old French *gromil* and the Modern French *Gremil.* Gromwell was reported to have been used by North American Indians as an oral contraceptive.

Ground-cherry, Clammy *Physalis heterophylla*

See Chinese Lantern Plant for the derivation of the generic name.

The Greek species name means "vari-leaved," a reference to the great variation in leaf shape. The resemblance of the berry to a yellow cherry and the plant's lowly stature explains the common name.

Ground-cherry, Virginia *P. virginiana*

See also Ground-cherry, Clammy.

This species has narrow, tapered leaves and yellow flowers with purplish spots. The bladder is deeply dented at the base.

The ripe berries of both this and the preceding species taste like tomatoes (to which they are allied botanically), and when cooked, they make a good preserve. This species is also related to the cultivated Chinese lantern plant, the potato, and the eggplant. The first described specimen was found in Virginia.

Ground Ivy. See Gill-over-the-Ground

Groundnut *Apios americana*

The generic name, Greek for "pear," refers to the shape of the tubers of this climbing vine, which bears fragrant maroon to chocolate-colored flowers.

Groundnut   The tubers, called "groundnuts," were eaten by

North American Indians and by rural folks in more recent times. Captain John Smith speaks of "grounds nut as big as Egges, and as good as Potatoes, and 40 on a string not two inches under ground." The tubers can be roasted, parboiled, or sliced thinly and fried like potatoes. The seeds, after cooking, can be eaten as beans.

Groundnut; Dwarf Ginseng *Panax trifolius*
. See Ginseng for the derivation of the generic name.
Trifolius, a Latin word, refers to the whorl of three leaves, each divided into three leaflets. The small edible bulb produced by this plant accounts for the first common name; its status as a minor version of the true ginseng explains the other common name.

Dwarf Ginseng

Ground Plum *Astragalus crassicarpus*
Astragalus comprises two Greek words meaning "star" and "milk." This name was bestowed on this member of the pea family because of an old belief that encouraging the spread of this plant on pasture land increased the milk yield of cows that ate it. In other words, the stars shone on milk production with this plant fodder. The Latin-Greek specific name meaning "thick-podded" is descriptive of these large pea-like pods. They are succulent and edible when young.
The short, thick pods of this species are almost plum-like in shape, hence the common name.

Groundsel, Common or Stinking *Senecio viscosus*
See Cineraria for the derivation of the generic name.
The specific Latin name, meaning "sticky," refers to the sticky, hairy stems and leaves. These are also malodorous, as suggested by the adjective in the common name.
Groundsel has an interesting derivation. It is based on two Old English words, *gundaes* and *welgae,* meaning "pus absorber." The chopped leaves were

Common Groundsel

commonly used in rural England to reduce abscesses.

Groundsel Bush *Baccharis halimifolia*
This genus was named in honor of Bacchus, god of wine, since the roots of an allied species were used to spice up wines in ancient Greece. The specific name indicates that the leaves resemble those of *halimum,* the rockrose of ancient Greece. Lost to history is the reason for the similarity in the common names of this and the common or stinking groundsel. There is no resemblance between the two.

Gumbo. See Okra

Harbinger-of-spring to Hyssop

Harbinger-of-spring *Erigenia bulbosa*
"Spring born," the Greek meaning of this generic name, alludes to its early spring flowering, as does the common name. *Bulbosa* tells us that its food is stored in a bulb.

Hardhack. See Steeplebush

Hardy Orange *Poncirus trifoliata*
The New Latin word *poncire,* a kind of citron, is the basis for this generic name. The leaves are divided into three leaflets, or trifoliate. This small, bitter-tasting but aromatic orange can be grown in the open in protected locations as far north as New York.

Harebell *Campanula rotundifolia*
See Bellflower for the derivation of the generic name.
The Latin species name means "round-leaved" and refers to the basal leaves. The later upper leaves are almost needle-like in shape.
Harebell suggests that this species grows in open fields frequented by hares. The violet-blue bells nod from stender stalks.

Hawkbit; Fall Dandelion *Leontodon autumnalis*
The ray flowers supposedly resemble "lion's teeth," hence the Greek-Latin generic name. This species, in contrast to others that are spring flowering, blooms in the early autumn, thus the Latin name *autumnalis.*
This flower closely resembles the common dande-

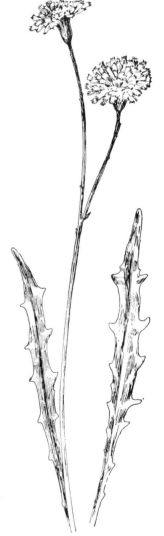

Hawkbit

131

lion and is in the same family. It also resembles the hawkweed, which is a partial basis for the common name. One dictionary referred to hawkbit as "a hawkweed with bitten roots."

Hawkweed. See Rattlesnake Weed
Hawkweed, Orange. See Devil's Paintbrush

Hawkweed, Rough *Hieracium scabrum*
See Devil's Paintbrush for the derivation of the generic name.

Scabrum, Latin for "rough," refers to the stem of this species. The common name is based upon the fable behind the generic name. The flower looks very much like a dandelion, but there are several on one tall stalk. Several species of hawkweed have come into cultivation.

Rough Hawkweed

Heal-all; Self-heal *Prunella vulgaris*
This generic name stems from the German *breaune,* meaning diseases of the jaw, mouth, or throat, such as quinsy or tonsillitis, for which this plant was believed to be a remedy. An earlier generic name, *Brunella,* strengthens the validity of the derivation suggested here. *Vulgaris,* Latin for "common," refers to the widespread occurrence of this species.

Both common names suggest the widespread belief in the healing properties of this lowly member of the mint family.

Heal-all

Heath, Spring *Erica carnea*
Erica springs from the Latin and Greek words for heath; *carnea,* Latin for "flesh-colored," refers to the color of the masses of small flowers. Heath derives from Middle English *heth,* or "waste land," the Anglo-Saxon *haeth,* and the German *heide.* Originally, it was applied to any plant growing on a heath or poor land.

Heath is related to the true heather. It grows no more than a foot tall and is used as a single specimen or for massed ground cover. It cannot stand excessive heat.

132

Heather; Scotch Heather *Calluna vulgaris*

Calluna derives from a Greek word meaning "to brush or sweep." This is based on the fact that at one time heather branches were widely used in making brooms. *Vulgaris,* meaning "common," attests to the wide distribution of this species. It has become naturalized in parts of the United States and is often planted in rock gardens and on dry slopes.

Heather is an ancient word, derived from the Middle English *hadder* and Norse *heithr.* It bears tiny, rosy-pink flowers in late sunmmer. White and crimson varieties have been developed.

Hedge Hyssop, Golden *Gratiola aurea*

This generic name is based on the Latin *gratia,* meaning "agreeableness" or "pleasantness," which describes the supposed medicinal virtues of this plant. *Aurea,* Latin for "golden," refers to the attractive yellow flowers of this swamp-loving plant. Many plants, this species among them, have been named hyssop because of a real or fancied resemblance to the aromatic herb of biblical times.

Hedge Mustard *Sisymbrium officinale*

An ancient Greek name of a fragrant herb, duly Latinized, was assigned to this genus. The Latin *officinale* refers to its former place in the market or herb shop. The common name attests to the fact that it is a mustard and that it often grows in the shelter of or near a hedgerow. This species has pungent seeds, and the leaves have antiscorbutic properties. The young leaves and shoots make a palatable potherb.

Hedge Nettle; Woundwort *Stachys palustris*

The dense spike of flowers is the basis for the generic name, the Greek word for "spike." *Palustris,* Latin for "of the swamp," is descriptive of the preferred habitat of the hedge nettle.

The common name is based on a superficial resemblance of the leaves to those of nettle, especially the sharp-toothed margins and prominent leaf veins. The name woundwort arose from the belief that this

Hedge Mustard

133

plant was useful in speeding the healing of wounds. Hedge nettle also was prescribed at one time as an antispasmodic and nauseant.

Hedge Nettle, Rough *S. tenuifolia*
See also Hedge Nettle.
The specific Latin name means "slender-leaved," a distinguishing feature. The stem is quite bristly, hence the common name. The rose-pink flowers are arranged in a circle at intervals around the stem, a unique feature of this plant.

Heliopsis; False Sunflower *Heliopsis helianthoides*
Heliopsis, Greek for "like the sun," and *helianthoides,* "resembling a sunflower," both allude to certain similarities, especially the yellow flowers which grow up to two and a half inches across. Double and long-blooming varieties are available to the gardener.

Heliotrope, Garden *Heliotropium arborescens*
Two Greek words for "turning toward the sun" make up this generic name. It is based on an old fable, since there is no evidence of such heliotropism in this genus. *Arborescens,* Latin for "shrubby," alludes to the compactness of this plant, which is usually treated as a half-hardy annual for summer bedding. Heliotrope has been described as a symbol of love and admiration. See also *Valerian.*

Helleborine *Epipactus helleborine*
Epipactus is an old Greek plant name later applied to this genus of orchids. This old name, in turn, was taken from the Greek word meaning "to coagulate," a reference to the use of the original plant in cheese-making.
The species name means "like a helleborus," the classical Greek name for Christmas rose. This native orchid bears an attractive spike of green flowers tinged with purple.

Hemlock, Parsley *Conioselinum chinense*

The Greek name for this genus is made up of two words meaning "dust celery." The application is obscure. The species name denotes the Chinese home of this introduced species.

The common name is based on the resemblance of the parsley hemlock and its leaves to parsley. This aromatic member of the parsley family is nonpoisonous and can be distinguished from poison hemlock by the lack of the telltale purple spots on the stem.

Hemlock, Poison *Conium maculatum*

The Greek name *conium* describes both the plant and its poison, which was administered to the condemned in ancient Athens. *Maculatum,* Latin for "spotted," refers to the purple spots on the stem, which distinguishes this species from the somewhat similar but nondeadly related plants.

One of a handful of poisonous plants, poison hemlock grows three to six feet tall, and the leaves, when bruised, emit an unpleasant odor.

Poison Hemlock

Hemp; Marijuana *Cannabis sativa*

Cannabis is the classical Latin name for this well-known drug plant; *sativa* attests to its long cultivation, now illegal in many countries.

It is commonly known as hemp because it has been grown for the hemp fibers in its stem. The second and better known name is a Spanish corruption of "Mary-Jane."

A native of Asia, marijuana has spread all over the United States except for the area from West Texas to California, Florida, and a strip along the border of Canada.

This plant is notorious as the source of a drug, and it is also the victim of a popular mix-up of sexes. The male plants, which die upon shedding their pollen, are called female hemp or femble; the female plants are referred to as male or carl hemp. The female plants live through early fall, until they are killed by frost.

Marijuana

Hemp Nettle

Hemp Nettle *Galeopsis tetrahit*
An ancient Greek word meaning "weasel-like" was chosen as the generic name because of the fancied likeness of the flower to the head of a weasel. The species name, Greek for "four-parted," refers to the four stamens in each flower. This coarse bristly herb superficially resemble the common nettle, hence its name.

Hempweed, Climbing *Mikania scandens*
This genus was named in tribute to Joseph G. Mikan, professor of botany at Prague University, in present-day Czechoslovakia, who wrote *A Catalogue of Plants According to the Linnaean System* in 1770. *Scandens,* Latin for "climbing," refers to the vining habit of this species. The composite flowers resemble those of boneset.

Hen and Chickens *Sempervivum tectorum*
The Latin generic name, translated "to live forever," refers to the tenacity to life displayed by this genus. The species name, Greek-Latin for "of the roof," denotes the fact that in humid climates hen and chickens often grows on roofs.
This plant produces a rosette of thick succulent leaves, from which smaller rosettes, or "chickens," are produced. The latter easily take root when separated from the "hen" or parent plant. This genus, in its many species, varieties, and hybrids, is popular in rockeries, garden edges, and on rock walls.
Among Pennsylvania Dutch, the sliced leaves are applied to warts, corns, and stings.

Henbane *Hyoscyamus niger*
"Pigbean" is the literal translation of this Greek generic name. The word "pig" is used in a derogatory sense as it is doubtful that swine ever ate the seeds of this plant, since it is known to poison poultry, livestock, and children. *Niger,* a Latin term, refers to the black color of these seeds, the source of the narcotic hyascamine.
As indicated above, this plant is "bane for hens."

Fowls that eat these seeds become paralyzed and die. Henbane has a disgusting odor and hairy stems.

Henbit; Dead Nettle *Lamium amplexicaule*
See Dead Nettle for the derivation of the generic name.
The specific name, Latin for "(leaves) clasping the stem," refers to a key characteristic of this plant. The seeds are eaten by birds and poultry, hence the common name, which means "a bit or morsel for the hen." *Dead nettle* signifies that this plant resembles a nettle but does not sting.
This mint, with round scalloped leaves and pur- Henbit plish-red flowers, can be used in spring as a salad and later as a potherb. Henbit juice, in small doses, has been used as a relief for asthmatic attack and whooping cough.

Henbit

Hepatica *Hepatica americana*
The generic and common names stem from the Latin and Greek word meaning "liver-like," which alludes to the liver-shaped leaves. The species name indicates that it is native to this continent.
Hepatica is one of the earliest spring flowers. The color varies from white and pink to lavender and blue. These are useful in shaded rock gardens.

Herb Robert *Geranium robertianum*
See Cranesbill for the derivation of the generic, species, and common names.
This strong-scented wild geranium bears small half-inch flowers and occurs in rocky woods in the northern half of the United States.

Hercules'-club; Devil's Walking-stick *Aralia spinosa*
Aralia is the Latinization of the common French-Canadian name *aralie*. The Latin word *spinosa* refers to the strong spines covering the stem. The two common names reflect the vicious club or cane, be- Herb Robert set with stout spines, that can be fashioned from the stout stem or trunk of this shrub.

Herb Robert

The bark produces a yellow essential oil used by Indians for decorative purposes. A closely related Japanese species is used as food. The very young leaves may be collected in the spring, cooked, and served with herb spices and vinegar.

Heron's-bill. See Stork's-bill
Hoary Puccoon. See Indian Paint

Hog Peanut

Hog Peanut *Amphicarpa bracteata*
The generic name, Greek for "double-fruited," refers to an interesting characteristic of the hog peanut. It bears two types of flowers, one showy which grows on the upper part of the plant. These form pods, each with three to four seeds. Inconspicuous, petalless flowers grow on creeping branches at the base of the plant. These form fleshy, one-seeded pods underground, sought after by hogs, hence the common name. The Latin species name refers to the bracts beneath the flowers. These beans make a palatable dish, if boiled and seasoned properly.

Hollyhock *Althaea rosea*
The hollyhock once had a use in medicine, since the Greek generic name means "that which heals." *Rosea* refers to its rose-colored flowers.
A native of China, the hollyhock reached England in the 1500s, where its alleged curative properties were appreciated. It gained the name hockleaf because the leaves were an essential ingredient in a concoction used to reduce the swelling of a horse's hock, or heel. Since many people had the mistaken notion that this new plant came from the Holy Land, it was named "holy hock" to distinguish it from other mallows.

Honesty; Moneyflower *Lunaria annua*
The round, silvery seed cases account for the generic name, Latin for "resembling the moon." *Annua* identifies it as an annual. Honesty, according to Asa Gray, can be an annual, biennial, or perennial,

138

depending upon where and how it is grown. Our observation is that it seldom flowers the first year, but produces abundantly the second year, and then dies.

Gray states that the name honesty came about because the seeds can be seen through the round pod. Gerard in his *Herball* says, "we call this herb in English Pennie flower . . . and among the women it is called Honestie." Other names are money-flower and dollar flower, all alluding to the large, round silvery pod, no thicker than a dime. Honesty sprays are very popular as a winter decoration.

Honewort *Cryptotaenia canadensis*

The hidden oil tubes which give honewort its chervil-like scent, are the basis of the generic name, which is Greek for "secret band or fillet." The species name indicates the Canadian origin of the first specimen to be described.

This root was popular in England in the seventeenth and eighteenth centuries for medicinal use in the reduction of a "hone," or swelling, hence the common name.

Today it is used to some extent as a seasoning similar to chervil; in spring and early summer it may be used as a salad, potherb, root vegetable, and soup flavoring. Honewort occurs in rich woods and along stream banks.

Honeysuckle, Fly *Lonicera canadensis*

This genus honors Adam Lonitzer (1528–1586), a German botanist and physician. He held the position of pensioned naturalist in Frankfurt, Germany, for 32 years. This afforded him the opportunity to observe and write extensively on botany and natural history. *Canadensis* refers to the locale of the specimen used to describe this species.

The common name arose from the fact that its nectar resembled honey and was sucked from the tubular flowers by flies and other insects.

This honeysuckle with yellow flowers has a very short flower tube, accessible only to certain flies.

Fly Honeysuckle

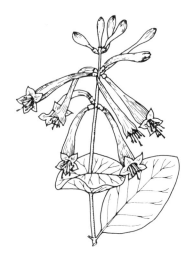

Trumpet Honeysuckle

Honeysuckle, Japanese *L. japonica*
See also Honeysuckle, Fly.
This is one of the most successful and aggressive of all plant immigrants. It quickly covers ground, bushes, and shrubs, smothering most plants in its path. Once established, it is very difficult to eradicate. Japonica refers to its original homeland.

Honeysuckle, Trumpet *L. sempervirens*
See also Honeysuckle, Fly.
The species name, Latin for "evergreen," applies only in the southern part of its range. The long tubular flowers are yellow within and red outside and occur in whorls. The roundish leaves are joined around the stem. This honeysuckle is scentless.

Hop, Common or European *Humulus lupulus*
Humulus is believed to be related to the Old Slavic *chumeli,* but basically is derived from the Latin *humus,* meaning "soil." This refers to its prostrate, humus-hugging habit when not supported. *Lupulus,* the Latin diminutive for "wolf," was the early generic name and is retained as a specific name. It alludes to the hop's ability to smother and kill the small trees and shrubs over which it grows very quickly—a botanical wolf, as it were.
Hop is a very old word, tracing to the Middle English *hoppe,* the Dutch *hop,* the Old German *hopfo,* and the modern German *hopfen.* The female plants produce the hops, a collection of cone-like fruits or seeds used in beer-making. Hops makes a good screen plant since a vine will readily grow thirty feet in a season. *H. aurea* is grown as an ornamental.

Horehound *Marrubium vulgare*
The ancient Hebrew word *marrob,* meaning "bitter juice," is the source of this generic term. It is descriptive of the bitter aromatic quality of the juice. *Vulgare,* Latin for "common," attests to the widespread occurrence of this species.

The common name stems from the Middle English and Anglo-Saxon *horehune,* "a gray plant," and alludes to its white woolly appearance. Horehound is best-known as a flavor for the familiar horehound candy. This mint was once popular as a stomach tonic.

Horn Poppy; Sea Poppy *Glaucium flavum*
The gray-green color of the leaves suggested the generic name, which is derived from the Greek word *glaukos.* Its flowers are yellow, hence the Latin adjective *flavum.*
The sickle-shaped seedpod, six to twelve inches in length, is the basis for the common name; its seashore habitat and poppy-like flowers explain the alternate name.

Horsebalm; Richweed *Collinsonia canadensis*
This genus honors Peter Collinson (1694–1768), an English botanist and friend of Linnaeus and Bartram. A self-educated Quaker, he became a noted gardener, whose gardens near London were famed for their exotics. There he introduced about 180 new species to England, 50 of which were from America. He persuaded American merchants and plant collectors to send him bulbs and seeds.

Mark Catesby, who collected in America from 1722 to 1726, was very helpful. Collinson gave Catesby an interest-free loan to enable him to complete his *Natural History of Carolina.*

John Bartram, another noted plant collector, aided Collinson. Collinson secured many wealthy customers for Bartram's seeds and bulbs. His horticultural work won him recognition by election as fellow of the Royal Society.

Collinson sent Franklin some glass tubes and details of German electrical experiments which got Franklin started in his study of electrical phenomena.

Horsebalm

This species was first described from a plant found in Canada, hence the species name.

The word "horse" was once used to describe anything coarse or strong; hence the name horsebalm, a balm fit for horses. The alternate name arose from the lemon-scented fragrance of its flowers. Horse-balm was once popular as a tonic, astringent, and diuretic.

Horse-gentian. See Tinkersweed

Horsemint *Monarda punctata*
See Beebalm for the derivation of the generic name.
The specific name, Latin for "dotted" or "spotted," refers to the purple spots on the yellow flowers. This plant was named horsemint because of its coarseness.

Horse Nettle; Buffalo Bur *Solanum carolinense*
See Bittersweet for the derivation of the generic name.
The species name means "from the Carolinas." The common names reflect the fact that this coarse prickly weed is widespread in fields and pastures, disliked by livestock, and a nuisance to man. The bright yellow berry-like fruit was once popular in the treatment of epilepsy.

Horseradish

Horseradish *Armoracia lapathifolia*
This generic designation is the classical Latin name of a related plant and later was assigned to this genus. "Sorrel-leaved" is the meaning of the Latin-derived species name. The "horse" part of the common name, used to describe something strong or coarse, refers to the strong pungency of this root. Horseradish is related to cabbage, turnip, mustard, and watercress. It often escapes from cultivation. It is widely used as a seasoning and is a reputed stomach stimulant, promoting secretions.

Horseweed *Erigeron canadensis*
See Fleabane, Daisy, for the derivation of the generic name.

Canadensis refers to Canada as the type locale for this species. Horseweed is a tall coarse weed, the "horse" being synonymous with coarse. It is common throughout North America. One plant can produce 20,000 to 40,000 seeds.

Hound's-tongue *Cynoglossum officinale*
See Comfrey, Wild, for the derivation of the generic name.

The species name, from Latin, signifies that this herb had a place in the herbalist's shop or in the market at one time. The common name is a translation of the Greek generic name.

Horseweed

The tiny flat seeds, covered with minute hooks, readily adhere to the clothing of a passersby. A number of varieties of hound's-tongue are under cultivation.

Huckleberry, Black *Gaylussacia baccata*
See Dangleberry for the derivation of the generic name.

The Latin species name means "berry-like or pulpy," an apt description.

The common name stems from the earlier English names, hurtleberry and whortleberry. This black edible berry has ten bony seeds which crack between the teeth, whereas blueberry seeds are so tiny as to be almost unnoticeable. See *Blueberry* for its uses.

Hyacinth *Hyacinthus orientalis*

Black Huckleberry

According to Greek mythology, Hyacinthus was in love with Apollo and was slain by Zephyrus. Hyacinthus' blood was changed to a beautiful flower which bore his name. *Orientalis* indicates its native habitat in the East, more specifically, in Asia Minor. This species is the ancestor of many cultivated forms.

The hyacinth was brought to England by Anthony Jenkinson, who went on a trading mission to Persia in 1561. Many new varieties were later brought in from Turkey.

This flower rapidly gained popularity in Europe, and by 1725 there were about 2,000 varieties under cultivation, largely in Holland. Favorite bulbs soon commanded high prices; in 1760 prize bulbs sold for £150–200 in England.

Hyacinth, Wild; Camass *Camassia scilloides*

The generic name and common name derive from *quamash,* the name given to this edible bulb by northwest American Indians. The Greek-Latin species name means "resembling a squill," which it does to some extent. The lavender and blue flowers are faintly suggestive of a hyacinth, hence the common name.

Hyacinth Bean *Dolichos lablab*

The ancient Greek name for the long-podded cow pea, based on the Greek word *dolichos,* or "long," was assigned to this bean at a later date. *Lablab* is based upon the original vernacular name in the East Indies.

The hyacinth bean bears showy white and purple flowers and produces a large purple pod. This ornamental has escaped from cultivation in many parts of southeastern United States. The tender young pods are edible when steamed; the beans may be eaten after cooking.

Hyacinth Bean

Hydrangea *Hydrangea arborescens*

The common and generic names are based on two Greek words meaning "water vessel," which describes the cup-form of the small seed vessel. The common name, Latin for "shrubby," is an apt description.

The small woodland shrub bears clusters of white flowers, the marginal ones being sterile. The cultivated hydrangeas, in contrast, have almost all sterile flowers in white, pink, and blue. A slight change in soil acidity will change a plant from pink to blue, and vice-versa.

Hydrangea

Hyssop *Hyssopus officinalis*

Hyssop, from the Hebrew *ezob,* is frequently mentioned in the Scriptures, though some authorities think the biblical hyssop was the caper. This Old World aromatic herb has become naturalized in many areas in the United States. The Latin specific name refers to its one-time importance in the market-place and in the apothecary's shop. Hyssop oil was used in liqueurs at one time.

Hyssop, Giant *Agastache foeniculum*
The dense flower spikes of this mint were fancied to resemble a wheat spike, the origin of the Greek generic name which means "resembling an ear of wheat." *Foeniculum* is Latin for "fennel-like," again a matter of resemblance.

The common name arose from the general simi-larity of this species to the related hyssop described above.

Ice Plant to Ivy

Ice Plant *Mesembryanthemum crystallinum*
"Midday flower" is the translation of the two Greek words from which the six-syllable generic name stems. It refers to flowers opening only in bright sunlight, about midday. The species name alludes to the crystalline appearance of the water vesicles in the leaves. These look like ice, hence the common name.

Impatience; Jewelweed; Snapweed; Touch-me-not
Impatiens pallida
The generic and common names all allude to the tendency of the ripe seedpods to explode on the slightest touch, scattering the seeds widely. *Pallida,* a Latin word, refers to the pale color of the flowers. This plant, also known as jewelweed for its delicate pendant flowers, was once reputed as a cure for poison ivy.

Indian Blanket. See Blanketflower
Indian Cucumber. See Cucumber Root

Impatience

Indian Cup; Painted Cup *Silphium perfoliatum*
This old Greek generic name was borrowed from that belonging to another resin-producing plant. *Perfoliatum,* a Latin word, describes the leaves as joined together around the stem. The conspicuous scarlet-tipped bracts around the small flowers suggested the common name. This is our native chewing gum plant. Country boys break off the flower stalks and return later to gather the blobs of hardened juice, which is a pleasant chew.

Indian Hemp. See Dogbane

Indian Paint; Hoary Puccoon *Lithospermum canescens*
See Gromwell, Corn, for the derivation of the generic name.

The specific name, Latin for "becoming white or hoary," refers to the foliage. Puccoon was the original Indian name of this plant. The common name derives from the Indians' use of the red-orange dye extracted from the roots.

Indian Paintbrush. See Indian Cup

Indian Physic *Gillenia trifoliata*
See False Ipecac for the derivation of the generic name.

Trifoliata, a Latin term, is descriptive of the three-part leaves of this herb. Its emetic roots were used by the Indians, hence the common name.

Indian Pipe *Monotropa uniflora*
Monotropa, Greek for "one turn," refers to the top of the flower stem, which turns to one side. *Uniflora,* meaning "one flower," indicates that each stalk bears but one flower.

This leafless saprophytic herb bears waxy-white nodding flowers, which turn black on withering. An extract of Indian pipe was once prescribed as a remedy for eye irritations.

Indian Pipe

Indian Plantain *Cacalia atriplicifolia*
Cacalia is an ancient plant name whose origin is lost in history. The specific name, meaning "atriplex-like leaves," is descriptive of the triangular lower leaves which resemble those of the saltbush (*Atriplex*).

147

Indian Tobacco *Lobelia inflata*

See Cardinal Flower for the derivation of the generic name.

Inflata, Latin for "swollen" or "puffed up," refers to the swollen calyx or flower base which becomes the seed capsule. This plant was often used as a tobacco substitute by the Indians, hence the common name. The leaves, similar to those of tobacco, may have prompted such use.

Ipecac, American *Gillenia stipulata*

See Indian Physic for the derivation of the generic name.

The stipules or leafy bracts at the leaf bases account for the specific name. The common name is based on the emetic property of this plant, similar to that of the Brazilian ipecac. This Tupi Indian word is translated as "small roadside emetic plant."

Indian Tobacco

Iris, Crested *Iris cristata*

See Blue Flag for the derivation of the generic name.

The iris became a symbol of Gaul in about the first century and a part of the French banner in the thirteenth century. The species name, from Latin *crista,* "a tuft," takes note of the prominent crest on the flower. The cultivated irises come in almost all colors of the rainbow and range from a few inches to three or four feet in height.

Ironweed, Tall *Vernonia altissima*

This genus commemorates William Vernon, a late seventeenth-century botanist and explorer who traveled in North America and made an extensive plant collection in Maryland in 1698. The common name alludes to the fact that this herb grows up to six or seven feet tall. It was named ironweed because of the hardness and stiffness of its stem.

Crested Iris

Ivy, English *Hedera helix*

Hedera is the ancient Latin name for ivy, *helix*

148

refers to the spiral manner in which some ivies climb. This Old World woody climber, naturalized in the United States, has evergreen, glossy leaves, which may be heart-shaped, kidney-shaped, or three-lobed, often with variegated patterns.

The word *ivy,* derived from the Greek, *iphyon,* has been altered to its present spelling as a result of passing through several languages and about two millenia.

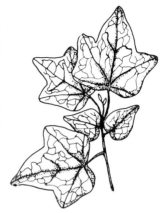

English Ivy

Jack-in-the-pulpit to Joseph's Coat

Jack-in-the-pulpit

Jack-in-the-pulpit *Arisaema triphyllum*
   See Green Dragon for the derivation of the generic name.
   Triphyllum, a Greek term, aptly describes jack-in-the-pulpit as having a leaf divided into three parts. The common name is based on the fancied resemblance of the club-like spadix to a preacher, Jack, standing in his pulpit, the spathe. This woodland denizen is also known as Indian turnip. The bulb contains crystals of a chemical which can severely burn the tongue. These are destroyed and dissipated when the bulb is thoroughly boiled. It is then edible —the Indian turnip. The jack, or spadix, bears staminate flowers in its early years and changes over to pistillate flowers in its third and later years. This sex change is known in several other perennial plants. Jack-in-the-pulpit occurs in green and purple phases.

Jacob's Ladder

Jacob's Ladder *Polemonium van bruntiae*
   There are two explanations of the origin of this generic name. The first ascribes its origin to the Greek word for "war," *polemos.* Pliny once wrote that a dispute over who discovered this pretty flower led to a brief war. Another account states that the flower was named in honor of Polemon, an early Athenian philosopher. The species name was given in 1892 in honor of Mrs. Cornelius van Brunt. She supplied specimens of this flower to the noted botanist, N. L. Britton, on the basis of which he described the new species in 1892, naming it in her honor. In the 1930s she substituted for a friend in the Research Department of the Brooklyn Botanic Garden.
   The common name is an allusion to the ladder on

150

which Jacob, in his dream, saw angels ascending into heaven. The long stem and pinnate leaves may have suggested a ladder to the imaginative person who first suggested this name.

Jade Plant *Crassula argentea*
The thick leaves of this plant explain its generic name, *Crassula,* from the Latin word for "thickish." *Argentea,* Latin for "silvery," refers to the general appearance of the plant. The common name is thought to refer to the color of the leaves of some varieties.

Jamestown-weed. See Jimsonweed

Jasmine *Jasminum sambac*
Eastern languages account for the origin of this name; the Arabic and Persian word *yasmin.* Sambac is an Arabic vernacular name.
Jasmine has been known in the Mediterranean area since ancient times, and it has been grown in England since the sixteenth century. It is cultivated for the beauty and fragrance of the flowers and also for the oil used in perfumery. Some species, such as *J. officinale,* are hardy as far north as Washington, D.C.

Jerusalem Cherry *Solanum pseudo-capsicum*
See Bittersweet for the derivation of the generic name.
The species name, part Greek, part Latin, means "false pepper" and relates to this plant sometimes being mistaken for a pepper plant.
The Jerusalem cherry is grown in pots for its attractive round scarlet fruits, which persist for a long time. A native of the Madeira Islands, this species is naturalized in Florida. The fruit resembles a cherry, and it was thought to have originated in the Holy Land, hence the "Jerusalem."

Jerusalem Cross. See Maltese Cross

Jewelweed. See Impatience

Jimsonweed

Jimsonweed; Jamestown Weed *Datura stramonium*
See Angel's Trumpet for the derivation of the generic name.

Stramonium is of Latin origin, its meaning unknown, and it is also the name of the dried leaves of the jimsonweed as used in pharmacy.

Also called Jamestown weed, it received its name from Virginia's first settlement, where the English colonists first encountered this poisonous, narcotic plant.

Robert Beverly, in his account of the early history of the Virginia colony, reports that in 1676 a body of soldiers was sent to Jamestown to suppress Bacon's Rebellion. Short of fresh rations, they collected young jimsonweed which they cooked as a potherb.

The consequence of eating this weed was reported by Beverly in these words:

> . . . they turn'd natural Fools upon it for Several Days: One would blow up a Feather in the Air; another woul'd dart Straws at it with much Fury; and another stark naked was sitting up in a Corner, like a Monkey, grinning and making Mows at them; a Fourth would fondly kiss, and paw his Companions, and snear in their Faces, with a Countenance more antick, than any in a Dutch Droll. In this frantick Condition they were confined, lest they should in their Folly destroy themselves; though it was observed, that all their Actions were full of Innocence and good Nature. Indeed, they were not very cleanly; for they would have wallow'd in their own Excrements, if they had not been prevented. A Thousand such simple Tricks they play'd, and after Eleven Days, return'd themselves again, not remembering any thing that had pass'd. (Robert Beverly, *History of Virginia,* p. 121).

The green prickly fruit, the size of a small apple, gave rise to another name, thorn apple.

Joe-Pye Weed *Eupatorium dubium*
See Boneset for the derivation of the generic name.

The specific name, Latin for "doubtful," suggests

doubt in the namer's mind as to its relationship to other species in the genus.

The common name recalls Joe Pye, an Indian herb doctor in the Massachusetts Bay Colony who claimed that a decoction of this weed cured typhus fever. The Iroquois used it as a remedy for kidney disorders, and the Chippewa drank a hot tea made of the dried leaves to produce sweat. The Meskwaki Indians regarded it as a powerful love medicine. A young brave who had a wad of leaves in his mouth while wooing an Indian maiden was assured of success in his mission.

Joe-Pye Weed

Jonquil. See Narcissus

Joseph's Coat *Amaranthus tricolor*
See Amaranth, Green, for the derivation of the generic name.

The blotched green, yellow, and scarlet leaves account for the species name and for the common name. This tropical annual is related to love-lies-bleeding and to prince's-feather.

Kalanchoe to Kudzu Vine

Kalanchoe *Kalanchoe pinnata*
Kalanchoe is based on the original Chinese name for this succulent perennial herb, popular during the Christmas season as a potted gift plant. The species name refers to the feathery leaves.
Another popular species, *K. coccinea,* bears long-lasting scarlet flowers.

Kale *Brassica oleracea acephala*
See Broccoli for the derivation of the generic name.
The Latin species name denotes it as a garden vegetable, and the varietal name is best translated as "headless cabbage." The common name is related to *cale* and *cole,* one of the cabbage tribe, and to cole slaw. This palatable winter potherb is closely allied to the wild cabbage of Europe.

Kangaroo Vine *Cissus antarctica*
Cissus, Greek for "ivy," was bestowed upon this climbing vine from Australia because its thick glossy foliage bore some resemblance to ivy. The species name indicates that in a botanical sense, the Antarctic includes a broad band of territory adjacent to the Antarctic Ocean. Kangaroo vine suggests that this vine was the haunt of climbing kangaroos.

Kerria, Garden *Kerria japonica*
This generic name honors William Kerr, an English gardener with a yen for travel. He introduced this and many other plants from China and Japan. The species name signifies Japan as the original habitat of the Kerria.

154

Single and double varieties are available from nurserymen, as is a double ball-like variety known as the globe-flower.

King Devil *Heiracium pratense*
See Devil's Paintbrush for the derivation of the generic name.
Pratense, Latin for "of the meadows," identifies the habitat of this weed. It reproduces by creeping stems and runners as well as by seed; a "devil" of a weed to eradicate.

Knawel *Scleranthus annuus*
The firm, hard bracts around the flower head inspired the generic name, which in Latin means "hard flowers." The species name identifies this as an annual. The common name derives from the German *knauel,* meaning "ball of thread," a name suggested by this wiry, bushy prostrate weed.

Knawel

Knotgrass. See Goose Grass

Knotweed, Prostrate *Polygonum aviculare*
See Buckwheat, Climbing False, for the derivation of the generic name.
The species name, Latin for "relating to birds," indicates that birds forage for knotweed seeds. The common name refers to the conspicuous, almost arthritic-looking joints at each branch of this lowly plant.

Kochia, Broom, or Summer Cypress *Kochia scoparia*
See Burningbush for the derivation of the generic name.
The Latin word *scoparia* describes the plant as "broom-like."
The common name alludes to its one-time use as a broom and to its feathery, cypress-like foliage.

Prostrate Knotweed

One annual variety, summer cypress, is cultivated for its feathery foliage, compact shape, and brilliant crimson color in the late summer. Asiatics use its young tips as a potherb and the seeds in bread-making.

Kohlrabi *Brassica oleracea caulo-rapa*
See Broccoli for the derivation of the generic name.

Oleracea identifies this as a garden vegetable, and the varietal name indicates a rape-like stem. The common name originated from the varietal name and from its name in Italian, *cavoli-rape*. The kohlrabi is like a turnip produced on a cabbage root, with leaves borne atop the tuber.

Krubi *Amorphophallus spp.*
An erotic imagination apparently contributed to the origin of this Greek generic name, which means "deformed phallus." It describes the form and shape of the tubers of this tropical Asiatic plant of the arum family. The common name was derived from the original vernacular name.

This funnel-shaped calla-like flower is chocolate or red-maroon in color and is grown in tubs, placed outdoors in summer. This is believed to be the largest flower in the plant kingdom. Several blooms in the New York Botanical Garden were 11 feet 3 inches long; the sheathe was 8 feet 10 inches.

Kudzu Vine *Pueraria lobata*
This Asiatic vine was named in honor of M. N. Puerari (1765–1845), a Danish botanist and professor of botany at Copenhagen University. Its lobed leaves account for the species name, and *kudzu* is a vernacular Japanese name.

This perennial vine, with tuberous starchy roots, is a vigorous grower; the stems grow 40 to 60 feet in a season and are capable of covering shrubs and small trees. The fleshy inner part of the root is eaten in Japan. *Kudzu* is used for erosion control and as hay and forage.

Labrador Tea to Lycoris

Labrador Tea *Ledum groenlandicum*
Ledon, the Greek name for the rockrose, was assigned to this genus for reasons that are not clear. The species name indicates that the first specimen was described from Greenland. This low shrub occurs throughout the Arctic region, from Greenland to Canada and south into the Appalachians.

The common name reflects the use of this plant as a substitute for India tea, especially in Labrador, beginning in the seventeenth century.

Ladies' Tresses; Nodding *Spiranthes cernua*
The Greek generic name, meaning "spiral-flowered," is descriptive of the twisted flower stalk of this orchid. The flowers arch downward, hence the specific name, which means "nodding."

Ladies' Tresses

The bestower of the common name fancied these delicate white flowers as resembling ladies' tresses.

Lady's Slipper *Cypripedium acaule*
The slipper-like shape of the pouch portion of this orchid is the basis for both the common and the Greek generic names. The latter is translated to "Venus' slipper." *Acaule*, meaning "stemless," refers to the stemless leaves. One of our most beautiful wild orchids, this species occurs in acid-soil woods, very often under mature stands or pine. An allied species, the showy lady's slipper or moccasin flower, is the state flower of Minnesota.

Lady's Slipper

Lady's Thumb *Polygonum persicaria*
See Buckwheat, Climbing False, for the derivation of the generic name.

Persicaria, Latin for "peach-like," refers to the resemblance of these leaves to those of the peach. The leaves often show a blackish blotch that looks vaguely like a lady's thumb print. The peppery foliage of this and allied species make an easily available camp seasoning. Some are very mild; others, extremely pungent.

Lamb's-quarters; Pigweed; Redroot *Chenopodium album*
Two authorities, at different times and places, fancied this leaf to resemble a goose foot and the quarter of a lamb. The former provided the generic name, Greek for "goose foot," the latter, the common name. Pigs were observed to relish this weed, hence the alternate common name.

Tender shoots of lamb's-quarters can be boiled and eaten like spinach. It occurs everywhere, in gardens, roadsides, and waste places, so one need not go far afield to find it.

Lamb's-quarters

Lamb's Succory *Arnoseria minima*
"Lamb's chicory" is the translation of the Greek generic name. *Minima,* meaning "least or smallest," reflects the fact that this plant is scarcely ever a foot high. The common name attests to the lamb's preference for this small herb, naturalized from the Old World.

Lantana *Lantana nana compacta*
This old Latin plant name once was applied to the viburnum and later was assigned as a generic name for this group. The two-word species name describes this flower as "dwarf" and "compact." Lantana produces heads of verbena-like flowers, much used as bedding plants, for borders, and in window boxes.

Larkspur, Rocket *Delphinium ajacis*
See also Larkspur, Tall.
The species name arose from the petal marks thought to resemble the letters AIAI, that is, ajax.

158

This annual garden larkspur from Europe bears flowers that range from white through blue and rose-pink through violent.

Larkspur, Tall *D. exaltatum*
The fancied resemblance of the partly opened buds to a dolphin's head accounts for the Greek generic name. *Exaltatum,* Latin for "very tall," alludes to the tall stalk of this plant, usually three to six feet high.

The common name refers to the prominent spur on each flower, like a lark's spur. About 60 species of larkspur are native to the north temperate zone. These are parents of most named varieties and hybrids.

Rocket Larkspur

Lavender, Common *Lavendula spica*
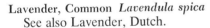
See also Lavender, Dutch.

Spica refers to the spikes of purple flowers. There are white and lilac-pink varieties in cultivation.

Lavender, Dutch *L. vera*
The generic name of lavender is based on the Latin word for "I wash." The Greeks and Romans used it in their baths, and today it is the basic ingredient in lavender water. The dried leaves are popular in sachets placed with stored clothing or linens and are used in perfumes to a limited extent.

Leadplant; False Indigo *Amorpha canescens*
This generic name, from the Greek word meaning "formless," refers to the small, imperfectly developed flowers, each with but one petal. The mass of tiny flowers form a dense violet spike. The specific name, Latin for "ashen or gray-white" is descriptive of the appearance of the hairy leaves.

Common Lavender

The leadplant was once believed to indicate the presence of lead deposits in the soil beneath its roots. Its other common name informs us that it was once the source of an indigo substitute.

Leafcup, Large-flowered; Bear's-foot *Polymnia uvedalia*

This genus was dedicated to Polyhymnia, the muse of sublime hymns or sacred songs. Note the word *hymn* in this name. The specific name honors Robert Uvedale (1642–1722), a horticulturist who grew this plant in his garden in England shortly after he received seeds from America.

The basal appendages of the leaves suggest a cup, hence the common name. The leaf shape—really more like a maple leaf—suggested the alternate common name.

Leatherflower *Clematis viorna*

See Clematis for the derivation of the generic name.

Viorna, meaning "traveler's joy," has an obscure application. Some Western Indians drank a tea made of an infusion of clematis bark to reduce fevers. This may have had a connection with the species name. The tea was considered an alterative, that is, it aided in restoring healthy bodily functions. The common name is descriptive of the leathery recurved petals of this wild flower.

Leek, Garden *Allium porrum*

See Chives for the derivation of the generic name.

Porrum is the classical Latin term for "leek." The work *leek* is related to the Anglo-Saxon *leac*, the German *lauch*, the Old Norse *lok*, and the Greek *lygos*. The latter meant a flexible slender twig, used as a band.

The leek has a long history of use among the peoples of Northern Europe. As the national emblem of Wales, it recalls the heroic resistance of the Britons against the Saxons in the year 640. St. David ordered the Britons to wear leek in their caps to distinguish them from the invaders. St. David's Day, March 1, is dedicated to the patron saint of the Welsh. An old couplet is recited, "March, various, fierce and wild with wind-cracked cheeks; By wilder Welshmen led, and crowned with leeks."

The bulbs are excellent as a base for soup or cut up into a tossed green salad.

Leek, Wild *A. tricoccum*

See also Leek, Garden.

The specific name means "three-seeded," a reference to the grouping of the seeds in threes. Leek was widely used by American Indians. They used it as a seasoning, for alleged aphrodisiacal properties, and to relieve pain of insect stings. The juice of crushed bulbs was rubbed on the skin for the latter purpose.

The Menominee Indians named a lakeside area, abundant with wild leeks, *shikako,* meaning a skunky or foul-smelling place. This was modified to Chicago, later the home of foul-smelling stockyards.

The young leek leaves, gathered in early spring while still unrolling, can be chopped and used in a salad. They also can be boiled in salt water and served with butter or sauce as a cooked vegetable. Use an onion or leek soup recipe in preparing the wild leek soup.

The cluster of greenish-white flowers appears on a single stalk in midsummer, long after the leaves have died down. If you spot the flowers, you will find the bulb delicacy a few inches below the soil.

Lemon, Ponderosa *Citrus limonia ponderosa*

This lemon, grown as an ornamental, takes its generic name from the classical Latin word for citron. The Latin species names mean "heavy" and "lemon."

Leopardflower. See Blackberry Lily

Leopard's-bane *Doronicum caucasicum*

The generic name stems from the vernacular Arabic name for this herb, *duronaj* or *doronigi.* Its original habitat was the Caucasus, hence the specific name. The common name arose out of a popular belief that this plant, grown in a garden or about the home, would ward off wild beasts. A relative of arnica, it is also known by the name of *Arnica acaulis.*

Letter-flower *Grammanthes dichotoma*

A V-like mark on the petals gave rise to the common name, which is also a translation of the Greek generic name. The twin-branched or forked habit of this plant led to the Greek specific name, related to the word *dichotomy.* This low annual, with fleshy leaves and yellow flowers, is grown in sunny rockeries.

Lettuce, Blue *Lactuca floridana*

See also Lettuce, Canada or Wild.

The specific name alludes to the locale from which the species was described. This blue-flowered lettuce grows three to six feet tall and has lyre-shaped leaves. It is also edible. The flowers are showier than those of other wild species.

Lettuce, Canada or Wild *L. canadensis*

Lactuca is derived from the Latin *lacta,* meaning "milk," as is the Old French *laitues* from which *lettuce* stems. This alludes to its milky juice. The species name refers to the locale of the first specimen described.

This common weed grows up to nine feet tall and bears heads of small yellow flowers. It is recognized in spring by its dandelion-like leaves and milky juice. It is a popular and nutritious salad and potherb, especially when young shoots are gathered early enough. By early summer it develops a bitterness which makes it suitable only as a cooked vegetable; use a little water and a pinch of salt. Vinegar, seasoning, and chopped onions add flavor to this tender potherb.

The author has extended the wild lettuce "season" by taking ten-inch tips of lettuce plants encountered while walking in a circle route. New side shoots were collected the following week when he retraced his steps. This went on for several cuttings. Wild lettuce can be frozen for future use in the way that spinach is frozen and stored.

Wild Lettuce

162

Lettuce, Prickly *L. scariola*
See also Lettuce, Canada or Wild.
The Latin species name refers to the thin, membranous leaves. The stem and mid-rib of the leaves are prickly, hence the common name. This lettuce grows three to five feet tall and bears yellow flower heads.

This is also known as compass plant, as the leaves turn during the day to keep one edge toward the sun. Still another name, opium lettuce, goes back to the nineteenth century when the milky juice was collected in late summer and coagulated into a rubber-like ball which looked and smelled like opium. At most it was a sedative.

Prickly Lettuce

This European species is believed to be the ancestor of the cultivated forms. Garden lettuce clearly discloses its ancestry when it is allowed to "bolt" in midsummer. Its flower stalk reveals the deep-lobed ancestral leaves, milky juice, and small dandelion-like flowers. Prickly lettuce is edible and can be prepared the way wild lettuce is.

Lettuce, White. See Rattlesnake Root

Lewisia *Lewisia tweedyi*
This genus commemorates two men who made important contributions to botany. Captain Meriwether Lewis was the leader of the famous Lewis and Clark expedition to the Pacific. He brought back a large collection of plant specimens, including some new ones. Frank Tweedy, a botanist, wrote the *Flora of Yellowstone National Park* in 1886.

This is considered one of the finest rock garden plants. It bears a rosette of fleshy leaves and rose or white flowers of waxy or satiny appearance. A related species, *L. rediviva,* also known as bitterroot, is the state flower of Montana.

Licorice, Wild *Galium lanceolatum*
See Bedstraw for the derivation of the generic name.
Lanceolatum, the Latin specific name, is descriptive of the lance-shaped leaves. The common name is traced through several languages: the Middle English

licoris, the Old French *licoresse,* and the Latin *liquiritia,* which was corrupted from the Greek *glykorrhiza,* meaning "sweet root."

Lilac

Lilac *Syringa vulgaris*
Syringa, meaning "tube" in Greek, refers to the tubular corolla of the individual lilac flower. *Vulgaris,* Latin for "common," attests to the widespread distribution of the lilac.

Lilac, an old English word, has its roots in the Arabic *laylak* and the Persian *nilak,* from *nil,* meaning "blue."

This fragrant shrub was brought to the American colonies before the 1700s. Since then over 530 varieties and 100 species have been introduced or developed in America. Lilac is the state flower of New Hampshire.

Lily, African; Lily-of-the-Nile *Agapanthus africanus*
"Flower of love" is the literal translation of this Greek generic name. There is no erotic connotation here; more likely, just a reference to a lovely flower. These showy perennials, with flowers in many shades of blue, are useful for terrace decoration in large pots. The species name indicates its African origin.

Lily, Amazon *Eucharis grandiflora*
The Greek root of this generic name means "pleasing" or "charming," an allusion to the fragrance and beauty of this lily. *Grandiflora* refers to the unusually large flower.

Lily, Butterfly *Hedychium coronarium*
The snow white and fragrant blooms of this lily suggested the generic name, made up of two Greek words meaning "sweet" and "snow." Its former use in garlands and wreathes gave rise to the species name, Latin for "wreath."

Lily, Foxtail; Desert-Candle *Eremurus himalaicus*
"Solitary tail," the translation of this Greek ge-

neric name, describes the appearance of the lengthy, single-flowered stalk. The specific name alludes to its original Himalayan habitat. This genus includes white, yellow, pink, and orange lilies, all native to West Asia and the Himalayas. Some produce an eight-foot flower stalk.

Lily, Kaffir *Clivia miniata*
The Duchess of Northumberland, a member of the noted Clive family in England, is commemorated in this generic name. The showy red flowers suggested the specific name, meaning "cinnabar red" in Latin. This lily originated in South Africa in the territory of the Kaffirs, a Bantu tribe.

Lily, Tiger *Lilium tigrinum*
Lilies have been under cultivation for more than 3,000 years. The name is very old, with roots going back to the Persian *lalek* and to the Greek *leiron*. Both the common and generic names trace to this origin.
The dark orange tiger lily, native to Japan and China, has rows of black spots that suggest a tiger's stripes. This lily produces a black bulblet at the base of each leaf. These bulbs and seeds are both used for propagation.

Lily, Turks-cap *Lilium superbum*
See Lily, Tiger, for the derivation of the generic name.
The species name attests to the striking appearance and elegance of this native lily. The stems, bearing several orange-yellow, black-spotted flowers, may be six or seven feet tall. The reflexed petals are suggestive of a Turk's cap, hence the common name.
This species is representative of the 85 species scattered over the north temperate zone. Most of these are under cultivation today.

Lily-of-Peru *Alstroemeria aurantica*
This genus commemorates Baron Claus Alstroe-

European Lily-of-the-Valley

Wild Lily-of-the-Valley

mer (1736–1794), a friend and pupil of Linnaeus. He made an official trip to Spain in 1753 to study sheep breeding, but he had Linnaeus in mind and sent him a collection of plants and seeds. Among these were seeds of a South American flower which later was named after him. Alstroemer continued his travels and plant collecting and published often in *Linnaea,* a botanical journal. *Aurantica* is Latin for "orange."

Several other species are in cultivation. Since alstroemerias belong to the amaryllis family, lily is somewhat of a misnomer.

Lily-of-the-Nile. See Lily, African

Lily-of-the-valley, European *Convallaria majalis*

This generic name stems from the Latin *convallis,* a "valley," a reference to the favored habitat of this plant. *Majalis* means "flowering in May." This is the fragrant, white-flowered perennial commonly grown in shaded borders or beneath shrubs around homes.

Lily-of-the-valley, Wild *Maianthemum canadense*

This genus was named in honor of Maia, mother of Mercury in Greek mythology, in whose honor the month of May was dedicated. The "flower of Maia" blooms in May. *Canadense* refers to the locale from which the first specimen was described.

Lily of the valley is a fitting name for this flower, which seeks out a shaded, moist vale as its home. Its white berries have a bittersweet taste and are somewhat cathartic. Unlike the European lily of the valley, this species bears a terminal spike of white flowers.

Lily-turf, Big Blue *Liriope muscari*

This genus was named after a Grecian nymph, Liriope. This small evergreen herb produces flowers resembling the grape hyacinth, hence the species name. The common name alludes to the grass-like foliage and to the lily-turf's use as a ground cover.

Lima Bean *Phaseolus lunatus macrocarpus*
See Bean, String, for the derivation of the generic name.

The two Latin descriptive words that follow the generic name mean "crescent-shaped" and "large-seeded," both apt for this bean. A native of the South American tropics, it is related to the civet bean. The common name is believed to pertain to Lima, Peru.

Lion's-foot; Rattlesnake-root *Prenanthes serpentaria*
The drooping flowerheads, characteristic of this genus, inspired the generic name, meaning "face-downward flowers." The common name was suggested by the leaf shape, and the alternate and specific names relate to the serpentine appearance of the root.

Lipstick Plant *Aeschynanthus lobbianus*
An imaginative botanist bestowed this generic name, two Greek words meaning "shame flower," an allusion to the crimson flowers he imagined to be blushing. This species name is unique in honoring brothers, William and Thomas Lobb, plant collectors for the James Veitch Nurseries, near Exeter, England. They are noted for the introduction into England of many new plants from Asia and South America. William went to California in 1849 and returned in 1853, bringing with him the first seeds of *Sequoia gigantea,* the giant redwood. Thomas collected extensively in East Asia, discovering and bringing back many new orchids. The common name stems from the red lipstick color of these flowers.

Lion's-foot

Live-forever; Orpine *Sedum telephium*
See Goldmoss for the derivation of the generic name.

The specific name honors Telephus, son of Hercules, for reasons lost to history. This plant holds on to life tenaciously, hence the common name. It will survive for months when suspended from a ceiling or lying on a table.

Orpine is an old name derived from the French

Live-forever

word *orpiment,* a soft yellow mineral pigment, and in turn from the Latin *auripigmentum,* a gold pigment. The English word dropped from five to two syllables while undergoing this change.

This garden and rockery plant is widely grown in the United States and frequently escapes from cultivation.

Lizard's-tail *Saururus cernuus*

The common name of this genus is a translation of its Greek generic name. *Cernuus,* meaning "drooping" or "nodding," is descriptive of the long flower stalk. This denizen of swamps and wet places bears a spike of feathery fragrant white flowers.

Lobelia, Great *Lobelia siphilitica*

See Cardinal Flower for the derivation of the generic name.

The species name *siphilitica* arose from the alleged efficacy of this plant in curing syphilis. This is the largest species of *Lobelia* and grows up to three feet tall. There are many smaller species, both wild and under cultivation.

Lizard's-tail

Loganberry. See Blackberry

Loosestrife, Fringed *Steironema ciliatum*

See Loosestrife, Purple, for the derivation of the common name.

"Sterile thread" is the translation of the Greek generic name. This refers to the sterile filaments which occur among the stamens in this species. The specific name calls attention to the fringe of tiny hairs, or cilia, near the base of each leaf.

Loosestrife, Purple *Lythrum salicaria*

This generic name, Greek for "blood," alludes either to the dark purple flowers of some species or to the reputed styptic quality of some species. *Salicaria,* Latin for "willow-like," is descriptive of the leaves. The common name is believed to be a mis-

Purple Loosestrife   168

translation of *lysimachia,* the original Greek name of this plant.

Lopseed *Phryma leptostachya*
The origin of the name of this genus is lost to history. The Greek specific name, however, is descriptive of the flower stalk, "slender-spiked." When the flowers have withered away, the ripe seeds are bent back or "looped" close against the long stem, hence the common name.

Lotus, American; Water Chinquapin *Nelumbo lutea*
Nelumbo, the Singhalese name for the water lily, was adopted as the generic name of our lotus. *Lutea,* "yellow," alludes to the huge yellow flowers, sometimes ten inches across.

Lotus is a classical Greek name; *water chinquapin,* an American Indian name, informs us that the immature seeds, boiled or roasted, taste like the diminutive relative of the chestnut. The "water" part of the name denotes this as an aquatic plant, denizen of sluggish streams and ponds.

The huge pepper-shaker seed capsule is easily recognized, as are the large floating leaves. These will lead one to the long rootstocks filled with starch in the autumn. When cut up and baked, they taste somewhat like sweet potatoes. In the spring leaf-stalks make a good potherb.

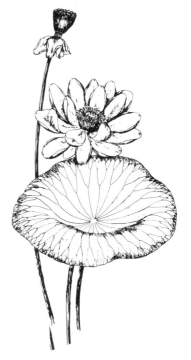

American Lotus

Lousewort *Pedicularis lanceolata*
"Lousy" is the simplest translation of this Latin generic name. This, and the common name, hark back to a bit of old folklore. It was once thought that the presence of this plant in a pasture encouraged the infestation of lice in sheep and cattle. *Lanceolata* refers to the lance-shaped leaves of this denizen of swamps and wet areas.

Lovage *Levisticum officinale*
This generic name is believed to be a corruption of *ligusticum,* the related Scotch lovage, which was named for its native Liguria, Italy, where it is abun-

dant. *Officinale* refers to its early status in the marketplace or herbalist's shop.

Lovage is a corruption of old French *luvesche,* which stems from the Latin *ligusticum* and *levisticum.* The root is reputed to have medicinal properties, and the flowering tops yield a volatile oil. Seeds and leaves are used for flavoring foods.

An allied species, the Scotch Lovage, *L. scothicum,* is established along the Atlantic coast from Labrador to New York.

Love-in-a-mist *Nigella damascena*

This generic name is a diminutive of the Latin word for black and alludes to the tiny black seeds. *Damascena* refers to Damascus, Syria, where this plant occurs. The common name is believed to be derived from the finely dissected floral bracts of the blue flowers.

This European garden flower, introduced into America, is related to black cumin, the condiment plant. Its aromatic seeds are used to flavor cheese, sausage, and other prepared foods.

Love-lies-bleeding *Amaranthus caudatus*

See Amaranth, Green, for the derivation of the generic name.

The beautiful red flower spike inspired the species name, which means "having a tail." The crimson chenille-like flower spike tops a stalk that is three to six feet tall. The romantic common name likewise is based on the crimson flower spike. This annual from India has several cultivated varieties.

Love-in-a-mist

Lupine, California *Lupinus polyphyllus*

See also Lupine, Wild.

Polyphyllus, a Greek word meaning "many-leaved," refers to the ten to sixteen leaflets into which each leaf is divided. This species is a parent of the popular Russell hybrids. It is stout and erect, growing two to five feet tall, with blue, white, or pink flowers. About 300 species of lupine have been described, and over 25 of these are in cultivation.

Lupine, Wild *L. perennis*
Both the generic and common names derive from *lupus,* Latin for "wolf." Centuries ago, it was believed that an invasion of pasture or cropland by this vigorous plant would deplete fertility and make the land useless, an effect similar to the inroads a pack of wolves would make in a herd of sheep. In reality we know that as a legume lupine fixes nitrogen in the soil, thus enhancing fertility. Lupine is not a botanical wolf! *Perennis* distinguishes this species as one which lives for many years rather than one or two. A southwestern lupine known as bluebonnet is the state flower of Texas.

Wild Lupine

Lychnis, Evening; White Campion *Lychnis alba*
Both the common and generic names derive from the Greek word *lychnos,* meaning "lamp." This name was inspired by the flame-colored, brilliant flowers of one or more European species, such as *L. chalcedonica,* the scarlet lychnis.

Our species is regarded as a weed, but many Old World species are under cultivation with quite an assortment of names: Jerusalem Cross, Mullein pink, Rose campion, Flower of Jove, to name but a few.

Lycoris *Lycoris squamigera*
This genus recalls Lycoris, a beautiful Roman actress and one-time mistress of Marc Antony. The species name refers to small scales in the throat of the flower tube.

Lycoris, a relative of the amaryllis, bears red, white, orange, and lavender flowers. The foliage appears first, then dies, and the fragrant flowers appear in late summer. It is also known as naked-lady and mystery-lady.

Evening Lychnis

Madder to Mustard

Madder, Blue Field *Sherardia arvensis*
This genus was named in honor of William Sherard (1659–1728), a noted English botanist. The species name, "of the cultivated field," indicates that it once had economic importance; it was used as a blue dye source.

The common name is related to several words all meaning "blue" or "blue dye": Danish *mede,* Polish *modry,* Old Norse *mathra.* It is related to and resembles the bedstraws.

Mahonia. See Oregon Holly-grape
Maize. See Corn, Sweet
Mallow, Common. See Cheeses

Mallow, Marsh *Althaea officinalis*
Certain mallows, once used in medicine, explain the generic name which is Greek for "that which heals." The species name, meaning "of the apothecary shop," lends support to the important place of mallow in the healing arts in the past.

Mallow, derivative of the Latin *malva* and the Greek *malakos,* "softening," refers to the emmolient quality of the sap. The juice from the root is the basis of a popular confection, the marshmallow.

Maltese Cross; Jerusalem Cross *Lychnis chalcedonica*
See Lychnis for the derivation of the generic name. This species is commonly found in Chalcedon, the classical name of an area near Istambul, Turkey. The common name was suggested by the shape of the flowers. Jerusalem cross is another popular name for this border perennial.

Mandrake. See Mayapple

Man of the Earth *Ipomoea pandurata*
Two Greek words meaning "similar to bindweed" make up this generic name. The reference is to the twining growth habit of *Ipomoea*. *Pandurata,* Latin for "like a fiddle," refers to the shape of the leaves.

The common name alludes to the tuberous roots which can grow to enormous size and often weigh twenty pounds (an undoubted relative of the sweet potato).

Maracock. See Passionflower

Marigold, African or French *Tagetes erecta*
This genus was named for the Etruscan diety, Tages, grandson of Jupiter, who sprang from the plowed earth in the form of a boy. *Erecta* refers to the upright posture of this species.

The common names are based on an erroneous belief as to their place of origin. This marigold was introduced into Europe by Charles V, who probably got it from Cortez, since *tagetes* is native to Mexico and South America.

Marigold, originally known as goldflower, became identified with the Virgin Mary and was considered a charm against evil power. The later name, Mary's Gold, was slurred into marigold.

Marigold, Marsh; Cowslip *Caltha palustris*
This generic name was borrowed from an earlier Latin name for a yellow-flowered plant. It derives from the Greek *kalathos,* "a goblet," a reference to the flower's shape. *Palustris,* Latin for "of the swamp," pinpoints this plant's usual habitat.

As for the common names, the *Caltha* is neither a marigold nor a cowslip. It is best described as a marsh-loving buttercup relative which bears over-size buttercup-like flowers.

Cowslip is from the Anglo-Saxon *cuslyppe,* a "cow's dropping," literally, "cow slop." The original

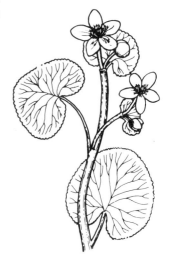

Marsh Marigold

cowslip of wet meadows grew best on sites of old dung deposits.

Marsh marigold leaves make an excellent pot-herb if they are boiled two or three times, a few minutes per boil, and the water is discarded each time. Similarly treated, the unopened flower buds can be pickled and used like capers.

Marigold, Water *Megalodonta beckii*

This is a sticktight gone partially aquatic. It has feathery submerged leaves and a few above-water leaves with large-toothed edges. The latter trait suggested the generic name, which is Greek for "big-toothed."

This interesting plant "amphibian" was named after its discoverer, Lewis C. Beck (1798–1853). He was a professor of botany at Rensselaer Polytechnic Institute and a professor of natural history at Rutgers University. He wrote *Botany of the United States North of Virginia* (1848). Its resemblance to the marigold, at least in color, and its aquatic habitat suggested the common name.

Marijuana. See Hemp

Mariposa Lily; Star Tulip *Calochortus venustus*

"Beautiful grass" is the translation of the Greek generic name of this western American bulbous plant. *Venustus,* meaning "handsome" or "charming," aptly describes these flowers. Most of the 40 to 50 species of *Calochortus,* all native to the west coast, are now in cultivation. Mariposa lily was named after the Spanish word for butterfly, "mariposa," an allusion to its colorfulness. Star tulip recognizes these as American relatives of the tulip of Asia.

Marjoram, Sweet *Origanum majorana*

This fragrant herb derives its name from Greek words meaning "delight of the mountain." This referred to a Grecian species commonly found growing on mountain slopes. Both the species and common names stem from the Arabic word "marjamie," the name of an aromatic herb. Marjoram is grown chiefly by herbalists and is used as a seasoning for soups, stews, meats, dressings, and coleslaw.

Marjoram, Wild *O. vulgare*
See also Marjoram, Sweet.
Vulgare, meaning "common," is a wild species, now often found as a roadside weed. It bears purplish-red flowers and is aromatic like sweet marjoram.

Marsh-pepper. See Smartweed

Marsh Pink; Rose Pink *Sabatia dodecandra*
This genus of gentian relatives was named for Liberato A. Sabati, a mid-eighteenth-century Italian botanist, author of several botanical works, and curator of the Rome Botanic Gardens. The species name refers to the "twelve stamens" in each flower. The common names allude to its habitat and color.

Marsh Pink

Matrimony Vine *Lycium vulgare*
Lycium, the old Greek name of another plant, is derived from Lycia, a former Roman province of southwest Asia Minor. *Vulgare,* Latin for "common," reflects the widespread occurrence of this vine.
According to an old tale, this plant was considered an evil eye for matrimonial bliss. If it thrived near a home, it foretold matrimonial misfortune. Possibly the thorny nature of the vine gave rise to this tale.

Mayapple; Mandrake *Podophyllum peltatum*
The large parasol-like leaf of the mandrake gave rise to the generic name, Greek for "foot leaf," which in turn is from an older word meaning "duck's-foot leaved." The incised leaf suggested this name.

175

The specific name, Greek for "shield-shaped" likewise applies to the leaf, taking the leafstalk as the shield handle.

The egg-shaped, edible fruit, resembling a lemon, ripens in late May in the South, hence the common name. The resemblance of the root to that of its European namesake gave it the name of mandrake.

The fruit makes good jelly and marmalade. A southern drink is made of mayapple juice, wine, and sugar. It is also mixed with fruit juice to make a novel summer drink. The root of mandrake has long been known for its cathartic properties and was so used in medicine.

Maypop. See Passionflower

Mayweed; Dog Fennel *Anthemis cotula*

Anthemis, based on the Greek word for "flower," refers to the free-flowering habit of this species. It was also the original Greek name of a medicinal herb. *Cotula,* Greek for "small cup," describes the fold at the leaf base.

Mayweed has strongly scented feathery leaves and is considered a weed. Its month of flowering gave it its common name, and its resemblance to true fennel accounts for the alternate name, that is, it was a false or "dog" fennel.

Meadow Beauty

Meadow Beauty; Deergrass *Rhexia virginica*

Rhexia, a classical plant name, is from a Greek word meaning "to rupture," which does not apply to this genus. *Virginica* refers to the state from which the first specimen was described. This pretty rose-colored flower, found in moist meadows, was aptly named meadow beauty. Its alternate name alludes to the fact that deer relish it. The leaves have a sweetish, slightly acid taste and make a pleasant salad.

Meadow Parsnip, Purple *Thaspium trifoliatum*

This name was borrowed from an allied genus, *Thaspia.* The stem leaves are divided into three

Purple Meadow Parsnip

176

leaflets, hence the Latin *trifoliatum.* This species has two varieties, one with purple flowers and a widespread variety with golden yellow flowers.

Meadow Rue *Thalictrum dioicum*

The ancient Greek name for another plant latterly was applied to this genus. The species name indicates that the male and female flowers are on separate plants. It is based on a Greek word meaning "two households." Rue refers to a herb with bitter-tasting leaves that one would "rue" (regret) sampling. This buttercup relative has greenish-white, purple-tipped flowers.

Meadowsweet *Spiraea latifolia*
The ancient use of these flowers in wreathes and garlands inspired the generic name, Greek for "garland." *Latifolia,* Latin for "broad-leaved," distinguishes the leaves from those of other spireas. Meadowsweet is an apt name for this denizen of damp meadows. It bears terminal clusters of pale pink to white flowers.

Meadow Rue

Meadowsweet, Aromatic *Filipendula vulgaris*
Filipendula, Latin for "hanging-thread," refers to the root tubers which are attached to thread-like roots. *Vulgaris* signifies "common" or widely distributed. The aromatic leaves of this relative of spirea are the basis for the common name. An oil, similar to oil of wintergreen, is distilled from the flowers, and the leaves are sometimes used to flavor soups.

Medic, Black *Medicago lupulina*
Medicago a classical name of a grass, later was applied to this genus. The name is from the Greek *medike,* "from Medea," the supposed country of origin. The specific name, meaning "hop-like," describes medic leaves. Another version of the derivation of the common and generic names is that they both stem from the Latin *medica,* "to heal," though there are no recorded medical uses of this plant.

Black Medic

Melon; Muskmelon *Cucumis melo*
See Cucumber for the derivation of the generic
name.
Melo and *melon* are from the Latin *melopepo,*
"apple-shaped melon," an appellation hardly applicable to today's melons. The prefix musk refers
to the sweet odor of melons.

Mercury, Three-seeded; Mercury-weed *Acalypha
virginica*
Acalypha, the ancient Greek name for nettle, was
later applied to this genus because of the nettle-like
appearance of the leaves. *Virginica* refers to the
state from which a specimen was first described.
This plant was named for Mercury, the ancient
messenger of the Gods, for reasons long lost to us.
It bears tiny green-brown flowers in the leaf axils.
Each flower produces three tiny seeds, the basis for
the "three-seeded" part of the name. This herb has
reputed diuretic and expectorant properties.

Mermaid Weed

Mermaid Weed *Proserpinaca palustris*
This aquatic herb was thought by the ancient
Greeks to be the abode of mermaids. Its generic
name, meaning "forward-creeping," refers to its
habit of growth. Its aquatic habitat is indicated by
the Latin *palustris,* "of the swamps."

Mignonette *Reseda odorata*
Reseda is from the Latin *resedere,* "to heal." The
name was bestowed by Pliny because of the plant's
use as a poultice for bruises and cuts. The magic in-
cantation "reseda morbos" was repeated many times
while the poultice effected its cure. This sweet-
scented (*odorata*) species is often planted by bee-
keepers as an important source of nectar for honey-
bees. The common name is a diminutive of the
French *mignon,* meaning delicate, dainty, and graceful, all appropriate descriptions of this species.

Milk Vetch *Astragalus canadensis*
An ancient belief held that encouraging this plant

in a pasture would improve a cow's or goat's milk yield. The generic name, Greek for "star" and "milk," somehow developed from this belief. *Canadensis,* "from Canada," refers to the locale from which this species was described. Milk vetch occurs over most of the eastern United States as well as Canada.

The common name relates to the folklore described above; that is, a vetch that promotes milk yield. The young pods of milk vetch are edible when boiled. In the west, related edible species are known as ground plum and Indian plum. These must be distinguished from the poisonous loco weed.

Milk Vetch

Milkweed, Common *Asclepias syriacus*
This genus was named by the Greeks in honor of Asclepias, god of medicine, since some species were highly regarded for their medicinal virtues. The specific name indicates that the American species resembles one from Syria. *Milkweed* refers to two attributes: milky juice and weed status.

Very young milkweed shoots can be cooked and eaten like asparagus. Similarly, the young pods can be boiled and served as a vegetable. If the flower clusters are cooked while still green and unopened, they resemble broccoli.

From Quebec Indians we learn of the use of milkweed as a contraceptive. An infusion made of the pounded roots produced temporary sterility in women. Note that these roots are considered poisonous by modern authorities.

Common Milkweed

Milkweed, Spider *Asclepiodora viridis*
"A gift of Asclepias" is the fanciful meaning of this generic name. *Viridis,* Latin for "green," refers to the predominant color of the flowers, which also have purple hoods. The origin of the descriptive "spider" is obscure.

Milkweed, Swamp *Asclepias incarnata*
See Milkweed, Common, for the derivation of the generic name.

The flesh-colored flowers of this swamp-loving

milkweed are the basis for the Latin specific name. This species may be eaten in the manner of the common milkweed.

Milkwort, Racemed *Polygala polygama*

This euphonious scientific name deals with lactation and marriage. In ancient times milkwort was believed to promote lactation. From this arose *polygala,* a Greek word meaning "much milk." *Polygama,* meaning "many marriages," refers to seed produced by closed flowers on underground branches. These are self-pollinated, a kind of botanical incest, and produce considerable seed. The normal aboveground flowers, usually cross-pollinated by insects, likewise produce seed. The common name also originated from the folklore described above.

Racemed Milkwort

Mint, Apple *Mentha rotundifolia*

According to a Greek legend, an unfortunate nymph, Mentha, was changed into an aromatic herb by the god Prosperpine because she incurred his displeasure. The Latin species name, meaning round-leaved, distinguishes this from other mints. The aroma is suggestive of apple flavor, hence the name. This European mint, one of the tallest of the genus, is locally naturalized in America and is frequently found in herb gardens.

Mint, Field *M. arvensis*

See Mint, Apple, for the derivation of the generic name.

The Latin word, *arvensis,* "of the cultivated field," explains both the specific and the common name's adjective. This mint is recognized by its tiny lilac flowers, densely crowded between stem and leaves.

Field Mint

Mistflower *Eupatorium coelestinum*

See Boneset for the derivation of the generic name.

The Latin species name, meaning "sky blue" aptly

describes these flowers. The flat flower head of tiny, crowded blooms suggested the common name.

Mistletoe *Phoradendron flavescens*
The Greek generic name, translated as "thief tree," is an appropriate description of this parasite on the branches of broad-leaved trees. Flavescens, Latin for "yellowish," refers to the thick, yellowish-green leaves.

The common name stems from an old belief that the mistletoe is propagated by bird droppings. It was centuries later that botanists noted that this parasite was spread by seeds accidentally sprouting from droppings or birds that had eaten the berries. The old Anglo-Saxon word for dung was "mist" or "mistle," and tan meant "twig." This was altered to *mistletoe* with the passage of time.

In the Middle Ages and later branches of mistletoe were hung from the ceiling to ward off evil spirits. Today, it is hung in the home purely as a Christmas holiday decoration. The berries are poisonous, if eaten.

Mistletoe is the state emblem of Oklahoma.

Mistletoe

Miterwort; Bishop's-cap *Mitella diphylla*
Mitella, Greek for "little cap," alludes to the form of the young seed capsule and its supposed resemblance to a bishop's miter or cap. *Diphylla* refers to the pairs of stalkless leaves. This moist woodland plant bears beautifully fringed white flowers resembling snowflakes.

Moccasin Flower. See Lady's Slipper

Mock Orange, Sweet *Philadelphus coronarius*
The ancient Greeks named a sweet-scented shrub in honor of Philadelphus, an Egyptian king who reigned from 285 to 247 B.C. The Latin word *coronarius* refers to the former use of the flowering branches in garlands.

This species, a garden favorite, bears large, fragrant single flowers. Hundreds of hybrids and double-

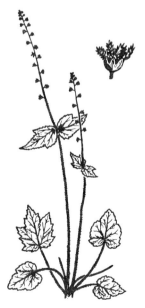

Miterwort

181

flowered varieties of mock orange have been developed from about 50 species.

Moneyflower. See Honesty

Moneywort

Moneywort *Lysimachia nummularia*
This genus was named by Dioscorides after King Lysimachus of Thracia. *Nummularia* is Latin for "resembling coins." The common name and its British synonym, herb twopence, are based upon the resemblance of the round leaves to a small coin.

Another explanation is that the generic name derives from two Greek words meaning "ending strife." Roman farmers are said to have put these flowers under the yokes of oxen to repel gnats and flies, thus sparing the animals the irritation of insect bites.

Monkeyflower *Mimulus ringens*
The likeness of this flower to a monkey accounts for both the common and generic names, the latter meaning "monkey" in Latin. *Ringens* refers to its "gaping" mouth. The lobed violet lips of this flower truly suggest a monkey's face.

Montbretia, Garden *Crocosmia aurea x pottsi*
This generic name is from the Greek *krokos,* "saffron," an allusion to the saffon odor of the dried leaves. This hybrid, a cross between the two species noted above, was created in 1880 in France.

A. F. C. de Montbret was the botanist on the French botanical expedition to Egypt in the eighteenth century. This iris relative produces orange to scarlet flowers.

Moonseed *Menispermum canadense*
Both the Greek generic name and the common name allude to the crescent-shaped stone seed in each black berry of the moonseed. These berries are bitter and poisonous and should not be mistaken for wild grapes. *Canadense* refers to the locale of the first specimen described.

Morning Glory, Ivyleaf *Ipomoea hederacea*
See Man of the Earth for the derivation of the
generic name.
Hederacea is Latin for "ivy-like leaves," true of
the leaf shape, but not its color or thickness. The
flowers are at their best in the morning, hence the
common name.

Morning Glory, Purple *I. purpurea*
See Man of the Earth for the derivation of the
generic name and Morning Glory, Ivyleaf, for the
common name.
This is a particular pest of cornfields and is found
throughout our area. The purple morning glory
originated as a cultivated form from tropical America, but it has escaped and become a weed everywhere.

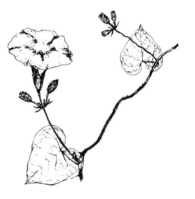

Purple Morning Glory

Moses in the Cradle *Rhoeo spathacea*
In Greek mythology, Rhoeo, the daughter of
Chrysothemis and Staphylus, was set adrift in a chest
by her father when he learned she had been seduced
by Apollo. This fable and the common name both
refer to the small flowers that develop in a boat- or
cradle-shaped structure arising in the angle between
the leaf and the stem. *Spathacea* refers to the spathe-
like structure around the flowers. The leaves of
Rhoeo, like the spiderwort to which they are related, are dark purple underneath.

Moss Pink *Phlox subulata*
The Greek word phlox, meaning "flame," alludes
to the bright flowers of some phloxes. The awl-
shaped leaves suggested the Latin specific name,
which means "like an awl." Moss pink is an apt
name for this lowly plant, which prefers rocks and
walls and bears masses of pink-like flowers.

Motherwort *Leonurus cardiaca*
Leonurus, Greek for "like a lion's tail," is a puz-
zler, since neither leaves nor flowers give us a clue to
the basis for this name. Even Asa Gray, father of

Moss Pink

Madder to Mustard

American botany, expressed puzzlement as to the reason for this name. Similarly, the species name, meaning "pertaining to the heart," cannot be explained.

Motherwort is from the Middle English *moderwort,* meaning "mother" and "root." This plant was once held to be valuable in treating diseases of the womb.

Motherwort

Mountain Laurel *Kalmia latifolia*

Peter Kalm, a Finnish pupil of Linnaeus, was sent by the Swedish government in 1748 to report on the natural resources of North America. He published a three-volume work on the natural history of North America and another work on the plants of Finland. Linnaeus named this genus in Kalm's honor. *Latifolia,* Latin for "wide leaf," distinguishes this from a related narrow-leaved species.

The evergreen, laurel-like leaves and rocky or mountainous habitat suggested the common name. Mountain Laurel is the state flower of Connecticut.

Mountain Mint *Pycnanthemum pilosum*

The Greek generic name, meaning "close flowers," refers to the dense clusters of tiny flowers. The Latin-derived *pilosum,* "covered with long, soft hairs," describes its leaves and stems. Despite its common name, this mint is found in fields, prairies, and dry woods, rarely in the mountains.

Mountain Laurel

Mousetail *Myosurus minimus*

The common and Greek generic names both describe the tapered flower stalk which ends up as a tail-like spike of seeds. *Minimus,* meaning "lesser," reflects the diminutive stature of mousetail, about six to eight inches tall.

Mugwort *Artemisia vulgaris*

See Dusty Miller for the derivation of the generic name.

Vulgaris, "common" in Latin, attests to the al-

most universal distribution of this aromatic perennial weed. Mugwort is derived from the Anglo-
Saxon *mucgwyrt,* a plant that repels gnats, flies, and
biting midges.

Mullein, Common *Verbascum thapsus*

The soft, dense, woolly leaves inspired the common name, from the Latin *mollis,* "soft." The generic name is an ancient one latterly transferred to
the mulleins. The specific name honors the town of
Thapsus in Sicily, or Thapsos in Greece, or the ancient African Thapsus (now Tunisia).

Roman soldiers, who manned forts in isolated
European outposts, made torches for lighting their
huts by dipping mullein into tallow. The thick down
is still used in isolated rural areas to make candle
wicks.

Mullein was introduced into America at an early
date, and the Indians soon learned of its value in
treating respiratory ailments, such as coughs, bron-
chitis, and asthma. They sought relief by smoking
mullein leaves. In the early 1900s a cough medicine
made of mullein flowers was very popular in America.

Mullein, Moth *V. blattaria*

See also Mullein, Common.

This common name arose from the fancied resemblance of the stamens and style of this flower to the
antennae and tongue of a moth. *Blattaria,* Latin for
"cockroach-like," offers no clue as to the basis for
this unusual specific name.

Moth Mullein

Mullein Foxglove *Dasistoma macrophyllum*

A Greek name meaning "woolly mouthed" was
bestowed on this genus in recognition of its trumpet-
shaped flowers with woolly petals inside. *Macro-
phyllum,* Greek for "large-leaved," aptly describes
the leaves.

This tall, yellow-flowered plant has some resemblance to both mullein and false foxglove, hence the
common name.

Musk Mallow

Madder to Mustard

Musk Hyacinth *Muscari moschatum*

See Grape Hyacinth for the derivation of the generic name.

Moschatum is derived from the Greek word for a musky, sweet scent, a characteristic of this species. This Old World cultivated species is planted in warm, dry borders. One stalk bears 20 to 50 tiny blue flowers.

Musk Mallow *Malva moschata*

See Cheeses for the derivation of the generic name.

Moschata refers to the musky fragrance of the foliage of this mallow, a garden form which often is found as an escape from cultivation.

Mustard, Black *Brassica nigra*

See Broccoli for the derivation of the generic name.

Nigra, Latin for "black," refers to the shiny black seeds of the mustard.

Mustard traces to the French *moustarde* and to the Latin *mustum,* meaning must. Long ago mustard was prepared by mixed ground seeds with some must, new grape juice before it begins to ferment.

Mustard has escaped widely from cultivation. Today it is grown commercially only in the Santa Clara area of California and in Montana. Mustard greens come from *B. japonica.*

Mustard, Hare's-ear *Conringia orientalis*

This genus commemorates Professor Hermann Conring, who held the chairs of natural philosophy, medicine, and law at the University of Helmstadt in Germany. He was physician to the Princess Regent of Friesland. The versatile Conring (d. 1681) wrote on natural history, medicine, chemistry, law, and history.

Orientalis, from Latin, refers to the Eastern (Old World) origin of this weed, now widespread in North America. The leaves resemble a hare's ear, hence the common name.

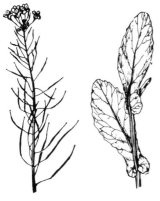

Hare's-ear Mustard

186

Mustard, Wild; Charlock *Brassica kaber*

See Broccoli for the derivation of the generic name.

Kaber is a Persian vernacular name for this mustard. A later synonym, *B. arvensis,* means "of the cultivated fields."

Charlock, an old term for this mustard, traces back to the Middle English *carlock* and to the Old English *cerlic.*

Wild Mustard

Narcissus to Norfolk Island Pine

Narcissus; Jonquil *Narcissus spp.*
Two divergent accounts are offered regarding the origin of this generic (and common) name. One account derives from the Greek word *narkeo,* "to be stupefied," a reference to bulbs of several species which possess narcotic alkaloids.

The second account is based on Greek mythology. Narcissus, pampered son of Echo, the river god, fell in love with a mountain nymph, but this love was not returned. Nemesis, god of vengeance, punished him for his persistence and egotism by transforming him into a flower. He would stand unto eternity at the edge of the river, nodding at his own image in the water.

There are over 10,000 named varieties of narcissus, daffodil, and jonquil, which are based on some 60 species native to Europe, Africa, and Asia.

Nasturtium *Tropaeolum spp.*
Linnaeus bestowed the generic name on this genus. He saw a likeness between the flowers and the Roman helmet or round shield, so he used the Greek word *tropaion,* meaning "trophy," as the basis for the name.

The common name *nasturtium,* "twist of the nose" in Latin, has a twisted history. This name originally was applied by the Romans to an evil-smelling watercress. When the nasturtium-to-be reached England from its native South America, it became popularly known as Indian cress because its pungent leaves were used in salads. In some manner it became confused with the Roman watercress and thereby acquired the name it now has. About 90 species of nasturtium are in cultivation, both annuals and perennials.

188

Nemesia *Nemesia strumosa*
Nemesia is an old plant name mentioned by Dioscorides and later applied to this group. There are over 50 species, both annuals and perennials, and their colors embrace virtually the entire spectrum. Many summer bedding varieties are derived from *N. strumosa*. The latter name means "having tubercles," a reference to the small sac or pouch at the base of the flower.

Nettle, Purple. See Dead Nettle

Nettle, Stinging *Urtica dioica*
Urtica is the original Latin name for nettle; *dioica,* Greek for "two households," indicates that male and female flowers are usually found on separate plants.
Watch out for nettle in moist, shaded woodlands. Leaves and stems are covered with stinging hairs. In spring and early summer, young shoots and leaves make a tender cooked vegetable, much like spinach. Gather them with gloves on!

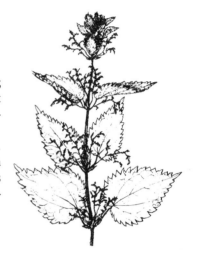

Night-blooming Cereus *Hylocereus triangularis*
The resemblance of cereus to the columnar shape of the wax candle and the plant's woodland habitat led to the generic name, Greek for "woods-wax-candle," and also led to its common name. *Triangularis* refers to the three-angled-stems. This unusual hardy cactus is a large-flowered nocturnal bloomer, climbs high via aerial roots, and bears very small spines.

Stinging Nettle

Nightshade, Black *Solanum nigrum*
See Bittersweet for the derivation of the generic name.
Nigrum refers to the black berries of this nightshade. They are believed to be poisonous, though they have been described as being used in cookery. Apparently they may be poisonous if eaten raw, but are safe when fully ripe berries are used in pies or preserves.

Black Nightshade

Nipplewort *Lapsana communis*

Lapsane, the ancient Greek name of this herb, is the antecedent of the generic name. The Latin specific name denotes that this is a species which grows in large communities. Nipplewort was once prescribed to heal nipple ulcers and was so used in Prussia, where it was called papillaris. This plant can be recognized by its yellow, dandelion-like flowers in midsummer.

Norfolk Island Pine *Araucaria excelsa*

Arauca, a locality in southern Chile where these pines grow, is the basis for the generic name. *Excelsa,* Latin for "tall," indicates the stature of this conifer in its native habitat. This pine is thought to have originated on tiny Norfolk Island in the South Pacific.

This tree is grown in a pot during its juvenile state and is used as an indoor or lawn decoration. It is unusually symmetrical and erect.

Ocean Spray to Oyster Plant

Ocean Spray *Holodiscus discolor*

A technical feature of this flower is the basis for its generic name, Greek for "entire-disked." The five minute ovaries are surrounded by an entire disk. *Discolor,* signifying "of two colors," is a reference to the flowers. Creamy white flower clusters on arching branches are suggestive of the ocean spray.

Oconee-bells *Shortia galacifolia*

This genus recalls Charles W. Short (1794–1863), a Kentucky botanist. Short combined a medical practice with indefatigable plant collecting. In a five-year period he distributed over 25,000 specimens to correspondents here and abroad. Add to this his own herbarium of 15,000 specimens, which he later gave to the Philadelphia Academy of Natural Sciences. Short served as professor of materia medica and medical botany at Transylvania University from 1825 to 1837.

The Latin species name refers to the "galax-like leaves" of this dwarf perennial from the mountains of the Carolinas. This and a Japanese species, *S. uniflora,* are used as ground cover for shaded, moist rockeries. The evergreen leaves are tinged with bronze. The common name alludes to the bell-shaped flowers, and the Oconee area of the Appalachians.

Oconee-bells has an interesting history of disappearance and rediscovery. It was first collected by André Michaux in 1788, and the specimen lay undescribed and unknown in a Paris herbarium until examined and described by Asa Gray in the early 1840s; he named it *Shortia.*

Botanists vied with one another to find living plants, both for herbaria and botanic gardens. Despite intensive efforts, Shortia was not rediscovered until 1877, 89 years after it was first collected! The next event was the discovery of a closely allied species in Japan in 1843, later named *S. uniflora*.

Oconee-bells is not classed as rare; it occurs fairly abundantly in a few localities. It can be grown as a pot plant or in heavy humus under mountain laurel, azalea, or rhododendron.

Okra; Gumbo *Hibiscus esculentus*

Hibiscus is the old Greek and Latin term for a mallow, which okra is. The Latin specific name denotes edibility. This vegetable was introduced from Africa into the West Indies and then into the southern United States. Both common names are corruptions of the original African vernacular words.

Okra is used in soups, stews, and catsup. It gives body to a dish because of its mucilaginous juices. Note the resemblance of okra to a hibiscus or mallow seedpod.

Oleander *Nerium oleander*

Nerium was the classical Greek name for this Mediterranean shrub. *Oleander* is from the Italian, *oleandro,* signifying, "olive-like," an allusion to the leaves. Oleander is grown outdoors in warmer parts of the United States for its attractive foliage and large leaves.

Onion, Garden *Allium cepa*

See Chives for the derivation of the generic name.

Cepa is the Latin name for onion, which in turn is derived from the Old French *oignon* and the Latin *unio,* "a single pearl," such as a pearl onion. Onion is believed to be a native of southwest Asia, but has been under cultivation for a millennium or longer throughout the civilized world.

Gerard, in his *Herball* writes that "the juice [of onion] anointed upon the pild or bald head in the sun, bringeth the haire againe speedily."

Onion, Nodding Wild *A. cernuum*
See Chives for the derivation of the generic name. The species name, meaning "drooping" or "nodding," identifies the posture of the flowers of this wild species.

Onion, Wild *A. stellatum*
See Chives for the derivation of the generic name. *Stellatum,* Latin for "starry," is descriptive of the showy umbel of lavender flowers of this species.

Orache; Saltbush *Atriplex patula*
Atriplex, the old Greek name for this plant, means, "not nourishing." Orache grows in saline soil, so the erroneous notion arose that this plant was not nourishing. As a potherb orache is superior to lamb's quarters. It is juicier and slightly salty. *Patula,* Latin for "slightly spreading," alludes to the leaf's lower lobes which spread out.
   The common name traces to the French *arroche* and the English-French *arasche.*

Orange-root. See Goldenseal

Orchid, Butterfly *Oncidium papilio*
The crest on the lip of this orchid looks like a tubercle or arrowpoint, which is the translation of the Greek generic name. *Papilio,* Latin for "butterfly," is an appropriate name for this reddish-brown and yellow orchid. This tropical American epiphyte produces flowers throughout the year, the basis for its popularity.

Orache

Orchid, Cranefly *Tipularia discolor*
This generic name stems from the Latin *tipula,* a "water spider," and the later word, *tipulid,* a "crane fly." The flowers are supposed to bear some resemblance to a crane fly. *Discolor,* Latin for "two-colored," refers to the two colors of this orchid. The

single purplish leaf persists through the winter and aids in finding this uncommon species.

Orchid, Purple-fringed *Habenaria blephariglottis*
See also Orchid, Small Woodland.
The pentasyllabic Greek specific name means "fringed-tongue," a reference to the lip of the flower. This bog orchid bears many white flowers on one stalk.

Orchid, Small Woodland *H. clavellata*
The long flower spur, shaped like a strap or rein, inspired the generic name, based on the Greek word *habena,* meaning "rein." This same spur suggested the specific name, which means "small, club-shaped" in Latin. Most American habenarias are found in bogs.

Small Woodland Orchid

Orchid, Showy *Orchis spectabilis*
Orchis, meaning "testicle" in Greek, refers to the resemblance of the bulbous part of the flower to the male organ. Its specific name, Latin for "showy," is well chosen for this flower with purple-rose hood and white lip and spur. Specimens are usually found singly in rich woods.

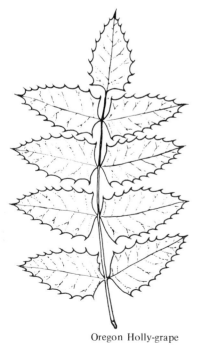

Oregon Holly-grape; Mahonia *Mahonia nervosa*
Bernard McMahon (1775–1816), an American botanist and horticulturist, wrote *The American Gardener's Calendar* (1807). He issued what is believed to be the first American seed catalog in 1804. The species name refers to the prominent nerves or veins on the leaves.
The common name stems from the grape-like berries and holly-like leaves of this low shrub, the state flower of Oregon. An ornamental, the Mahonia bears bright yellow flowers and dark blue berries. An alternate host for the deadly black stem wheat rust, it cannot be planted within several miles of a wheat field.

Oregon Holly-grape 194

Orpine. See Live-forever
Oswego Tea. See Beebalm

Oxalis; Wood Sorrel *Oxalis europaea*
The mild acid taste of the leaves of wood sorrel suggested its generic name, from Greek *oxys,* meaning "acid." The species designation refers to its European origin, though today it is found around the world.
The alternate common name, from Old French, means "little sour" plant, an allusion to the sour or acidly flavor of the entire plant. The leaves make an interesting addition to a salad, but they must be used in moderation since they contain oxalic acid.

Oxeye *Heliopsis helianthoides*
See *Heliopsis* for the derivation of the generic name.
The species name emphasizes the thrust of the generic name with a Greek word meaning, "resembling a sunflower." The center of this flower was fancied to look like an ox eye, hence the common name.

Oxeye Daisy *Chrysanthemum leucanthemum*
This white daisy with a yellow disk carries an appelation which recognizes both colors. The generic name is Greek for "golden flower," and the specific name means "white flower." See *Oxeye* for the derivation of the common name.
The name *daisy* can be traced through the Middle English *daiseie* or *dayesye* to the Anglo-Saxon *daegeseage,* that is, "day's eye." This refers to the bright "eye" or center disk of the flower. This large white daisy is found throughout the United States in fields and along roadsides. It is the state flower of North Carolina.
The young shoots of oxeye daisy are used as a salad in Europe and China, but the odor appears to discourage its use in America.

Oxeye Daisy

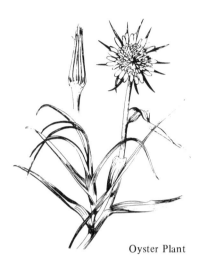

Oyster Plant

Oyster Plant; Salsify *Tragopogon porrifolius*
See Goatsbeard, yellow, for the derivation of the generic name.

The species name, Latin for "with leek-like leaves," is descriptive of the long, grass-like leaves of this plant.

The oyster-like flavor of the roots and leaves led to its popular name. The alternate name derives from the French *salsifis,* the Italian *sassefrica,* and the Latin, *saxifraga,* to which this plant is supposed to bear resemblance.

The young stems and lower leaf bases are used as a cooked vegetable; the boiled roots are eaten much like the cultivated variety.

Pachysandra to Quince

Pachysandra *Pachysandra terminalis*
The thick stamens noted in the white flowers of this ground cover plant led to the generic name, which in Greek mean just that. The specific name refers to the characteristic terminal flower stalk.

Pachysandra

Painted Cup; Indian Paintbrush *Castilleja coccinea*
See also Indian Cup.

This western plant was named in honor of an eighteenth-century botanist, D. Castillejo, of Cadiz, Spain. *Coccinea,* Latin for "scarlet," denotes the color of the flower bracts. This is the state flower of Wyoming.

The attractive scarlet-tipped bracts beneath each flower suggested both common names. The Navajos used a decoction of painted cup flowers as an application to painful centipede and insect bites.

Painted Tongue. See Velvet Flower

Pansy, Garden *Viola tricolor*
See Violet for the derivation of the generic name.

Tricolor identifies the pansy as a three-colored flower, the parent of many of today's cultivated varieties.

In the seventeenth and eighteenth centuries, the pansy was worn by unmarried French ladies of fashion in a bouquet, often with marigolds. The bouquet served as a delicate invitation to "think of me" or "remember me." The French called it *pensée,* "thought." When introduced into England at a later date, the name for this floral bouquet was changed to *pansy.*

Painted Cup

Pansy, Wild or Field *Viola kitaibelliana*

See Pansy, Garden, for the derivation of the common name.

This species honors Paul Kitaibel (1757–1817), professor of botany and chemistry at the University of Budapest, Hungary, and Director of the Botanical Garden in that city. He was author of scores of papers and several books on botany. This European pansy has been naturalized in this country.

Parsley, Garden *Petroselinum hortense*

A Greek word meaning "rock-parsley" was chosen as the generic name for this herb, whereas the specific name from Latin means "of the garden," a recognition of its cultivated status.

The English, noted for their facility in condensing words, reduced the pentasyllabic generic name to the two-syllable common name, a change that occurred slowly over several centuries. From the Greek we have the Anglo-Saxon *petersilie*, the Middle English *persely,* to the modern *parsley.* This plant, useful in flavoring soups, meat, fish, and salads, often runs wild in fields, near gardens, and along roadsides.

Parsnip, Garden *Pastinaca sativa*

Parsnip is corrupted from the Middle English *passenep* and ultimately derives from the Latin *pastinaca,* meaning "parsnip," and from *pastinum,* "to dig and trench the ground."

Sativa is Latin for "sown in a garden and field." This spring and winter root crop, native of Europe and Asia, was cultivated in Europe before the Christian era. It often runs wild as a pesty weed.

Parsnip, Water *Sium suave*

See Skirret for the derivation of the generic name.

Suave, Latin for "sweet," refers to the sweet root, edible when cooked. The common name is based on a resemblance to a parsnip plant.

Partridgeberry *Mitchella repens*

The partridgeberry or checkerberry was named in

honor of John Mitchell (1680–1768), who was a
friend of Linnaeus and Franklin and who is believed
to be the author of *American Husbandry.* A resi-
dent of Middlesex County, Virginia, he also was a
mapmaker and served as justice of the peace while
practicing medicine.

Upon his return to England in 1746, he prepared
a map of British and French North America, the
most important one of the period. He also introduced many North American plants to Britain and
described these in botanical publications. He developed a method of treating yellow fever that saved
thousands of lives in the Philadelphia epidemic of
1793.

The Latin word *repens* refers to the creeping or
trailing habit of this plant. The common name is
based on the observation that the scarlet berries were
relished by partridges.

Pregnant Cherokee and Penobscot Indian women
drank a tea made of partridgeberry leaves for a few
weeks prior to delivery. This tea was believed to
hasten labor and childbirth. This tea was listed in
the National Formulary from 1926 to 1947 for its
use as an astringent and diuretic.

Partridgeberry

Partridge Pea *Cassia fasciculata*

Cassia is the Latinization of the Greek name for
the biblical plant, *kasian,* to which partridge pea is
closely allied. *Fasciculata,* Latin for "grouped together in bundles," apparently refers to the drooping
anthers. Partridges are believed to have sought out
the pea-like pods of this plant, hence the common
name. The leaflets are sensitive to touch and fold
up after such contact.

Pasqueflower *Anemone patens*

Anemone, from the Greek word for "wind," refers
to the windswept habitat of the original plant. *Patens,*
Latin for "spreading," alludes to the sepals which
spread backward.

Pasqueflower, or "Easter flower," originally was
the French word *passefleur,* but was altered by Gerard
to its present name. *Pasque* is believed to be derived from the Hebrew *pesah,* to "pass-over," an

Partridge Pea

199

allusion to the biblical story of the Passover.
This prairie denizen has five to seven large, white
to purple petal-like sepals. These blossoms were
stuffed into the nose to stop nosebleed by Western
Indians.

Pasqueflower, European *A. pulsatilla*

See Pasqueflower for the derivation of the generic and common names.

Pulsatilla was the original sixteenth-century generic name of the pasqueflower. It is derived from a
Latin word, "to shake or sway," as in the wind.

The pasqueflower, bearing blue and purple flowers, appears about Eastertime, hence its common
name. The juice makes a green dye used to color
Easter eggs. The American pasqueflower is the state
flower of South Dakota.

Passionflower; Maypops; Maracock *Passiflora incarnata*

Passionflower

This flower received its Latin generic and its primary common name from early missionaries in
South America who saw the story of the crucifixion
in the flower parts. The corona was the crown of
thorns, the five sepals and five petals made up ten
of the apostles (omitting Peter and Judas). Other
flower parts represented nails and wounds. *Incarnata,* Latin for flesh-colored, refers to one of
the several colors of this flower.

The two common names stem from the Algonquian Indian words for the plant and the lemon-like
and edible fruit. The Indians believed that maypops
would relieve insomnia and soothe nerves. The fruit
has a pleasant taste and is eaten by rural people in
the South. Passionflower is the state flower of Ten-
nessee.

Pea, Garden *Pisum sativum*

Pisum, the Latin and Greek name for the pea, is
also the origin of the common name, traced through
the Middle English *pese,* the Anglo-Saxon *pise,* to
the original *pisum. Sativum* means "cultivated in
gardens and fields." One of the oldest cultivated
plants, the pea was known long before the Christian

era. There are three main types today: smooth, wrinkled, and edible pod.

Pea, Sweet *Lathyrus odoratus*
See Beach Pea for the derivation of the generic name.
Odoratus, from Latin, refers to the sweet perfume of these pea flowers. This popular flower garden plant bears white and rose flowers.

Peacock Plant *Calathea makoyana*
See Calathea for the derivation of the generic name.
Jacob Makoy, a prominent Liège, Belgium, horticulturist, introduced the peacock plant in 1872. It had been sent to him from the jungles of Minas Geraes in Brazil. The name was suggested by the striking variegated foliage, which has markings of olive green, cream, dark green, and red.

Peanut; Groundnut *Arachis hypogaea*
This generic name is a contraction of the earlier Greek *arachidna,* a clover which buried its seed head beneath the soil, as the peanut does. The Greek species name, meaning "underground," describes the unusual trait of the female flower. After fertilization the flower stalk sheds its petals, and turning earthward, it buries itself in the soil, where the peanuts develop.
The "nut" of the peanut resembles a large pea with husk removed, hence the common name. The derivation of *Groundnut* is obvious.
The peanut is grown for pods, forage, and hay from Virginia southward, but it can be grown anywhere as a garden novelty.

Pearly Everlasting *Anaphalis margaritacea*
The classical Greek name for another everlasting, *anaphalis,* was latterly applied to this plant. The pearly white flower head suggested the common name and species name, which is Greek for "pearl-bearing." This is the showiest of the everlastings,

Pearly Everlasting

with a globular flower head. Male and female flowers are on separate plants.

Pencil Flower *Stylosanthes biflora*

This flower has a long, hollow, pencil-like base. Hence the common name and the Greek generic name, which is translated to "pillar-flower," the Greek approximation of the common name. The orange and yellow flowers usually occur in pairs, hence "biflora." Another common name, afterbirth weed, is based on this plant's reputed usefulness as a uterine sedative.

Pencil Flower

Pennycress, Field *Thlaspi arvense*

Thlaspi is the ancient Greek name for cress. The word *arvense,* "of the cultivated field," indicates its preferred habitat. The flat, round pods are suggestive of the old English penny.

The young leaves of this cress, which have a mustard-onion flavor, are used in salads in the spring, and the seeds are used as a seasoning.

Pennyroyal, American *Hedeoma pulegioides*

The sweet aromatic odor of this herb gave rise to its generic name, derived from two Greek words meaning "sweet aroma." The species name translates to "resembling a fleabane," which it does to some extent.

The common name has no connection with pennies or royalty; it is a corruption of *pulyole ryale,* from the Old French *pulial real* and eventually from the Latin *pulegium,* meaning "fleabane."

Pennyroyal tea was used by American Indians to induce abortions and to relieve headaches and menstrual cramps. It was listed in the "U.S. Pharmacopoeia" from 1831 to 1916. Pennyroyal oil is an effective mosquito repellant.

Pennywort, Water *Hydrocotyle americana*

Field Pennycress

Two Greek words meaning "water" and "small cup" make up this generic name. The latter pertains to the cup-shape form of the leaves. *Americana* re-

202

fers to its New World habitat.

Pennywort, an old plant name, is descriptive of the rounded leaves, resembling a penny. It was once used in the treatment of skin diseases. Young plants and shoots of a related species are popular in salads and as a potherb in Asia, but it apparently has gone unnoticed here.

Peony *Paeonia officinalis*

This genus is named for Paeon, a physician in ancient Greece and pupil of Aesculapius. Plato is said to have turned Paeon into a medicinal herb, the first to be used in medicine. *Officinalis* indicates that this plant was sold in the marketplace or apothecary's shop. This European species is a parent of many cultivated forms. Another species, *P. albiflora* (meaning white-flowered), also is an ancestor of many hybrid peonies.

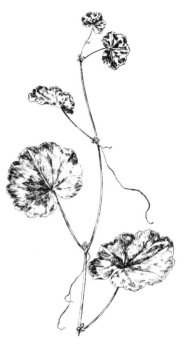

Water Pennywort

Peperomia *Peperomia marmorata*

The Greek name of this genus, which means "pepper-like," describes the decorative appearance of this foliage plant. *Marmorata,* Latin for "marbled," refers to the mottled or marbled appearance of the foliage.

Pepper *Capsicum grossum*

Two versions are extant regarding the origin of *Capsicum.* One is the Latin *capsa,* a "box or chest," suggested by the dry pepper with seeds. The other is the Greek word *kapto,* "to bite," an allusion to the pungency of pepper. *Grossum,* Latin for "thick," refers to the thick walls of the pepper.

The common name, shared by related species, can be traced through Anglo-Saxon, Latin, and Greek to the Sanskrit *pippali.*

A native of the New World, the pepper was brought to Europe by Columbus. His fleet physician lists it as a condiment (1494) and refers to its seeds as good bird feed.

Pepper has varied uses: as a salad or cooked vegetable, the source of paprika, a basic ingredient of chili con carne. Cayenne pepper is made from a related species, and still other varieties are grown as ornamental plants.

Peppergrass *Lepidium virginicum*
Lepidium is a classical Greek name, derived from two words meaning "small scale." This refers to the small, flat, peppery pods which look like scales. *Virginicum* refers to the state from which the species was described.

The common name aptly describes the peppery taste of the seeds. This little herb is used as a seasoning for salads and soups. In the spring the young shoots can be used like watercress in salads. A good meat dressing is made from peppergrass seeds, vinegar, and salt. A brew made of the tops of *Lepidium* was once popular in the treatment of dysentery and sore throat.

Peppermint *Mentha piperita*
See Mint, Apple, for the derivation of the generic name.

Piperita means "with a pepper-like aroma," which doesn't quite hit the mark.

Peppermint leaves or the essential oil are used to flavor sweets, chewing gum, and liqueur. It makes a refreshing hot or cold tea. See *Pepper* for derivation of common name.

Pepper-root. See Toothwort

Peppermint

Pepperweed, Field *Lepidium campestre*
See Peppergrass for the derivation of the generic name.

Campestre, Latin for "of the fields," refers to a common habitat of this European species, now widely naturalized here. A close relative of peppergrass and used in much the same way, it is distinguished by its large, clasping leaves.

Perilla; Beefsteak Plant *Perilla frutescens*
The generic and common names are derived from

Field Pepperweed

a vernacular Hindu name. *Frutescens,* Latin for
"shrubby or bushy," describes this compact plant.
The large red leaves suggested the alternate name.
This cultivated plant often escapes from cultivation.

Periwinkle *Vinca minor*
Vinca, the classical Latin name for this popular
ground cover, is a contraction of the Latin plant
name, *pervinca,* the origin of the common name.
This blue-flowered creeper is a frequent escape from
cultivation.

Periwinkle, Madagascar or Cape *V. rosea.*
See Periwinkle for the derivation of the generic
name.
The species name refers to the color of the flower
of this species. This popular window box and bedding plant blooms throughout the summer and bears
rose, purple, and white flowers. A common type is
V. rosea var. oculata, white-flowered with a pink
"eye." Periwinkles are easily started from seeds
planted indoors early in spring.
The eminent plant encyclopedist L. H. Bailey
states that there is no clear evidence that this periwinkle originated in either Madagascar or in the
cape area of South Africa, though it is of tropical
origin.

Periwinkle

Petunia *Petunia integrifolia*
Petunia is the Latinized version of the aboriginal
Brazilian name for tobacco, *petun,* a close relative.
The species name is Latin for "entire-leaved." This
purple species, together with *P. violacea* and *P.
nyctaginiflora,* both from Argentina, are the parents
of most hybrids now in cultivation.

Philodendron, Heartleaf *Philodendron cordatum*
The *philodendron* climbs trees in its native tropical American habitat, hence its name, which is
Greek for "tree-loving." *Cordatum,* from the Latin,
alludes to its "heart-shaped" leaves. This popular

house and display plant is valued because of its large
and variegated foliage and its calla-like flowers.

Phlox, Blue *Phlox divaricata*
See Moss Pink for the derivation of the generic
name.
Divaricata, Latin for "widespreading," alludes to
the growth habit of this species.

Blue Phlox

Pickerelweed *Pontederia cordata*
This genus commemorates J. Pontedera, professor
of botany at Padua University in Italy. *Cordata,*
from the Latin, is descriptive of the heart-shaped
leaves. This herb grows in the shallow waters of
streams and ponds, sharing its habitat with pickerels.
The starchy, nut-like seeds are pleasant eating, and
the young leaf stalks can be used as a potherb.

Piemaker. See Velvetleaf
Pigweed. See Lamb's Quarters

Piggyback Plant *Tolmiea menziesii*
This plant was named in honor of William F.
Tolmie (1812–1886), a Scottish physician who went
to British Columbia to become surgeon for the Hudson Bay Company at Fort Vancouver, B.C. He later
became chief factor and served until his retirement
in 1870. He was interested in the botany of British
Columbia and collected extensively. Tolmie made
the first recorded ascent of Mount Rainier. The species name recalls Archibald Menzies, a fellow surgeon and naturalist. The common name describes
the plantlets which form at the apex of the leaf stalk,
which, on falling, readily establish themselves nearby.
Tolmiea is used in borders and in wildflower gardens, especially in the Pacific Northwest.

Pickerelweed

Pigweed; Redroot *Amaranthus retroflexus*
See Amaranth for the derivation of the generic
name.
The Latin species name describes the unique man-

ner in which the dense spikes of greenish flowers bend or curve backward. This herb was relished by pigs, hence the common name.

In this genus, the ornamental species are known as amaranths; the weeds, as pigweeds and other earthy names. This species can be used like spinach as a potherb, especially in spring and early summer.

Pigweed

Pimpernel, Scarlet *Anagallis arvensis*

An ancient fable held that this plant had the power to dissipate sadness by generating "laughter" and "delight," the meaning of the Greek generic name. *Arvensis* "of cultivated fields," denotes a common habitat of this plant.

Pimpernel is from the Old French *pimpernelle,* the Spanish *pimpinela,* and the Latin *pipinella.* The latter, in turn, comes from *bipinnula,* the diminutive of the word for "two-winged." This refers to the feathery leaves of the plant which originally bore this name. Pimpernel is used as a potherb in Europe, but it is apparently untried here.

Pimpernel, Yellow *Taenidia integerrima*

Taenidia is Greek for "narrow fillet," a term which appears to fit the narrow leaflets of this pimpernel. The specific name, Latin for "most entire," refers to the entire leaflets, that is, they are not toothed or lobed.

See Pimpernel, Scarlet, for the derivation of the common name.

Pinesap *Monotropa hypopithys*

See Indian Pipe for the derivation of the generic name.

The species name, Greek for "under fir trees," refers to the habitat, but it is more often found under pine trees. This parasitic plant takes its nourishment from the sap of pine roots close to the surface.

Pinweed *Lechea minor*

This genus honors Johann Leche, an eighteenth-century Swedish botanist. *Minor* refers to the lowly

Pinesap

stature of this species. Pinweeds have short, grass-like leaves, hence the name. This herb was once popular as a tonic, fever reducer, and antiperiodic.

Pipewort *Eriocaulon septangulare*

The very woolly stem of this bog plant caught the eye of the bestower of the generic name, Greek for "woolly stem." The usually seven-angled flower stalk accounts for the Latin specific name. The pipewort was so named because of the membranous tube surrounding the ovary of this flower.

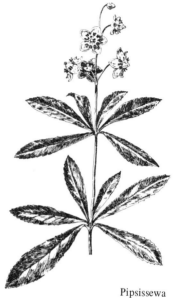

Pipsissewa *Chimaphila umbellata*

This Greek-derived generic name signifies "lover of winter," an allusion to the evergreen foliage, conspicuous in the winter. The terminal clusters of waxy-white flowers form an umbel or "little umbrella," hence the Latin species name.

Pipsissewa is Cree Indian for "juice breaks stone in bladder into small pieces," implying a belief in the efficacy of pipsissewa in treatment of bladder- or kidney-stones. They also used the astringent leaves as a tonic and diuretic.

The Pennsylvania Germans used pipsissewa tea to induce sweating. It once was a regular ingredient of root beer.

Pipsissewa

Pitcher Plant *Sarracenia purpurea*
J. A. Sarrasin, a sevententh-century French botanist and physician and a member of the French Academy of Sciences, was sent to Quebec as royal physician. There he discovered this carnivorous plant, later named in his honor. He was first to send this unusual plant to Europe. *Purpurea,* Latin for "purple," refers to the coloration of the pitcher.

Called pitcher plants because the hollow leaves resemble pitchers, the leaves are adapted to the capture and absorption of insects which venture into the trap at the mouth of the "pitcher" leaf. Bristly, down-pointing hairs prevent escape, and the luckless insect is slowly absorbed in a enzymous pool at the base of the hollow leaf. Only its outer skeleton remains as witness to its demise.

Pitcher Plant

Pittosporum, Japanese *Pittosporum tobira*
  This Greek generic name means "pitch-seed," a reference to the resinous coating of the seeds. *Tobira* is the Japanese vernacular name. This winter-flowering shrub with thick, leathery leaves and white or yellow fragrant flowers is grown out of doors in Florida and California, and indoors elsewhere.

Pittosporum

Plantain, Bottle-brush *Plantago aristata*
  See also Plantain, Common.
  Aristata, Latin for "slender spine-like tip," describes the seeds which are bristled like oats. This plantain was used by early settlers as a brush in cleaning bottles.

Plantain, Common *P. major*
  Both the generic and common names derive from the Latin *plantago,* meaning "sole of the foot," a name based on the shape of the leaf. *Major,* meaning "large," refers to its large leaves.
  Plantain was used in colonial medicine as an antidote for insect and snake bites. In more recent times the early spring leaves have been used in salads and as a cooked vegetable.

Plantain Lily *Hosta albo-marginata*
  Nicholas T. Host (1761–1834), physician to the emperor of Austria, was also a noted botanist. When not ministering to the health needs of the emperor and his family, he found time to write three notable monographs: "Plants of Austria," "Description of Grasses of Austria," and Volume 1 of *Salix* (Willow tribe). The species name from Latin refers to its white-margined leaves. The common name alludes to its plantain-like leaves. Several species of hosta are in cultivation.

Bottle-brush Plaintain

Pogonia, Rose; Snakemouth *Pogonia ophioglossoides*
  The conspicuous bearded lower lip of this rose-colored orchid accounts for the Greek generic name. The six-syllable species name, Greek for "like a snake's tongue," and the alternate common name

Rose Pogonia

209

describe the appearance of the flower. This attractive, sweet-scented orchid, is found in wet meadows and bogs.

Pogonia, Whorled or Five-leaved *Isotria verticillata*
See Pogonia, Rose, for the derivation of the common name.

Isotria, a Greek word meaning "in equal threes," refers to the three attractive madder-purple sepals. The whorl of five stemless leaves atop the stalk accounts for the Latin species name, which means "whorled."

Whorled Pogonia

Poinsettia; Christmas Flower *Euphorbia pulcherrima*
See Crown of Thorns for the derivation of the generic name.

The species name, Latin for "most beautiful," is an appropriate choice for the poinsettia. The upper leaves and bracts are a bright vermilion red. The tiny greenish flowers, discernable only to the botanically trained, consist of one central female flower surrounded by numerous male flowers.

The common name commemorates Joel R. Poinsette (1775–1851), gardener, botanist, and first U.S. Ambassador to Mexico in 1824. He introduced this Mexican plant to South Carolina, where it rapidly won wide popularity. Since it normally blooms in December, it has become a favorite Christmas gift plant.

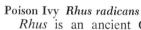

Poison Ivy *Rhus radicans*
Rhus is an ancient Greek name of this genus. *Radicans,* a Latin term signifying "rooting," refers to the black aerial roots by which poison ivy clings to tree trunks. All parts of the plant are poisonous to the skin, causing itching blisters. The plant is recognized by its shiny ivy-like leaves, each divided into three leaflets, and dirty-white seeds. The woody stems can be erect and shrubby or a vine that climbs high into trees. A decoction of the leaves is reputed to be a diuretic and nervous-system stimulant.

Poison Ivy

Pokeweed; Puccoon *Phytolacca americana*
The generic name, Greek for "lac plant," requires an involved explanation. Linnaeus, who named this plant, was impressed with the purple staining qualities of the poke berries, which he compared to the dark red dye extracted from the lac aphids of Asia. These feed in large numbers on certain trees, and the resin they deposit on twigs is also used in making varnish and wax. *Americana* signifies the American habitat of this weed.

Poke, or Pokeweed, is derived from the Algonquian Indian *puccoon,* "a plant used for staining or dyeing." The young poke leaves and sprouts are unexcelled as a spring salad and as a cooked vegetable, the latter comparable to asparagus. The young shoots also can be pickled. Poke becomes bitter and inedible by early summer.

Pokeweed

Pond Lily. See Spatterdock, Small

Pondweed *Potamogeton americanus*
The Greek words for "river dweller" were chosen as the generic name for this aquatic plant, which bears both floating and submerged leaves. There are 38 species of this genus in North America. The rootstock is edible when roasted or boiled. A rake and a rowboat are useful in harvesting a root crop.

Poorweed *Diodea teres*
The Greek word for "thoroughfare" was bestowed as the generic name, since this weedy plant is so often found along roadsides. The specific name, from the Latin word meaning "rounded," describes the round cross-section of the stem. The common name is believed to have arisen because this plant thrives in poor soil.

Poppy, California *Eschscholtzia californica*
This genus recalls the work of Johann F. von Eschscholtz, an early nineteenth century German naturalist and professor of anatomy and medicine at

California Poppy

211

Dorpat University in Russia. He traveled extensively, collecting plant and animal specimens in California, the Aleutian Islands, South America, and on many Pacific islands. He discovered and described many new species, including the poppy genus named for him. A bay in Kotzebue Sound, Alaska, also bears his name.

The specific name identifies the original habitat of this popular species. The bright-colored, poppy-like flowers suggested the common name of the popular annual. This is the state flower of California.

Poppy, Oriental *Papaver orientale*
Papaver is the classical Greek name for poppy, originally derived from *pap,* meaning "milky juice." *Orientale* refers to the Asiatic origin of this popular garden perennial. About 100 species of poppy are known, many under cultivation.

Portulaca; Moss Rose *Portulaca grandiflora*
Linnaeus, who named this genus, noted that the lid of the seed capsule opened like a gate. So the Latin word *portula,* "little gate," came to mind, and a new generic name was born. *Grandiflora,* Latin for "large-flowered," is appropriate for this South American species, widely cultivated in the United States. It bears white, orange, yellow, or red flowers.

Potato *Solanum tuberosum*
See Bittersweet for the derivation of the generic name.

The specific name, Latin for "tuber-bearing," refers to the discovery of this tuber, perhaps a thousand years ago, by the Indians of Central and South America. The potato was brought to Spain in the mid-sixteenth century, and Sir Francis Drake introduced it to England in 1586. A Royal Society committee recommended it for cultivation in Ireland in 1663 as a safeguard against famine. It soon acquired the name Irish potato. A potato blight devastated Ireland in 1846, resulting in the mass emigration of Irish to America.

Breeding has brought about great changes in this tuber. Heriot, reporting on the resources of colonial Virginia, describes the roots "as large as walnuts, hanging together as if tied on ropes."

Potato is derived from the Indian *batata* and the Spanish *patata*. The potato is closely allied to tomato and eggplant.

Pot of Gold; Golden Coreopsis *Coreopsis tinctoria*

See Coreopsis for the derivation of the generic name.

The species name, Latin for "of the dyer's," refers to the one-time importance of coreopsis as a dye plant. The abundance of bright gold-yellow flowers suggested the common name.

Pothos *Scindapsus pictus*

An old Greek plant name meaning "ivy-like tree" was applied to this genus of climbing vines. The much variegated leaves suggested the species name, which is Latin for "painted." Pothos is the vernacular Ceylonese name for this tropical houseplant, which resembles philodendron.

Pot Marigold. See Calendula

Pouch-flower; Lady's Slipper *Calceolaria integrifolia*

This large attractive flower which resembles a slipper or pouch prompted the generic name, based on the Latin *calceolus* or "slipper." The Latin species name is descriptive of the "entire," that is, unlobed, leaves.

Prayer Plant *Maranta leuconeura*

This genus was named in honor of Bartolommeo Maranta, a sixteenth-century Venetian botanist and physician who apparently devoted more of his time to botany than to medicine. He was director of the Naples Botanical Garden in Italy, 1554–1556, and developed a botanical garden in Rome in 1568. He grew rare and odd herbs and shrubs and wrote a

213

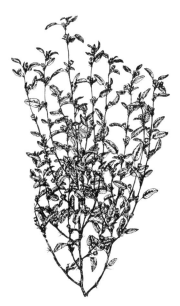

three-volume work entitled *Methods of Recognizing Medicinal Plants* (1559). He found time to serve as attending physician to Cardinal Castiglioni for many years.

Leuconeura, Greek for "white-nerved," refers to the narrow white bands along the leaf veins. The leaves of maranta tend to fold upward at night, like a pair of hands in prayer, hence the common name. This species is grown for its attractive foliage.

Prickly Mallow; False Mallow *Sida spinosa*
Sida is an ancient Greek plant name mentioned by Theophrastus in his writings. The spines at the bases of the arrow-shaped leaves explain the specific name. The yellow flowers have a slight resemblance to mallow, hence the common names.

Prickly Mallow

Prickly Saltwort. See Russian Thistle

Primrose, Evening *Oenothera biennis*
This genus has the unlikedly Greek name of "wine-imbibing." An infusion of the roots of one species of evening primrose was supposed to be an incentive for wine drinking or an enhancement of one's capacity to imbibe. The Latin species name denoted this as a biennial, that is, the plant grows the first year, develops a strong root system, and flowers and dies the following year.

The common name is based on two characteristics: first, a slight resemblance to the primrose, and second, the opening of the flowers as evening approaches and their wilting the next day.

The first year roots, dug in the fall or winter, are edible when cooked. The taste is betwen salsify and parsnip.

Primrose, Garden *Primula auricula*
Both the generic and common names have the same origin. Chaucer's *primerole* was preceded by the French *primeverole,* meaning the "first flower of spring." This name was first applied to another early

Evening Primrose

214

bloomer and later transferred to the *Primula. Auricula,* an old English name for primrose, is based on the Latin word meaning "ear-shaped," a reference to the leaf.

Primrose-willow. See Water Primrose

Prince's Feather *Polygonum orientale*
See Buckwheat, False, for the derivation of the generic name.
This species originated in India, hence the Latin name, meaning "from the Orient." The showy, drooping crimson spikes of small flowers gave rise to the common name in England in the seventeenth century.

Privet, Common *Ligustrum vulgare*
Ligustrum was the classical Latin name for privet. *Vulgare* refers to its common occurrence. *Privet* is an old English name for this hedge plant, which belongs to the olive family.

Puccoon, Hoary *Lithospermum canescens*
See Gromwell, Corn, for the derivation of the generic name.
Canescens, Latin for "ashy-gray," is an apt description of this hoary, hairy plant. *Puccoon* is the Algonquian Indian name for a plant which yields a yellow or red pigment. Puccoon root yields a red dye, which the Indians used for decorative purposes. They also are reported to have used the plant in the preparation of an oral contraceptive.

Common Privet

Pumpkin, Garden *Cucurbita pepo*
See Gourd for the derivation of the generic name.
Pepo and the earlier Greek *pepon* were the collective names for pumpkins and related cucurbits. The pumpkin had been cultivated and used by the American Indians for many centuries before the European discovery of America.

Puncture Vine *Tribulus terrestris*

This Latin generic name, meaning "three-pointed," refers to the number of sharp prickles usually on a bur of the seedpod. *Terrestris,* from Latin, signifies "hugging the ground" and is descriptive of this plant.

The thick seedpod contains five burs, each with two to four sharp, stout spines, strong enough to puncture a bicycle tire or the sole of a lady's shoe.

Puncture Vine

Purslane *Portulaca oleracea*

See Portulaca for the derivation of the generic name.

Oleracea refers to its usefulness as a potherb. The mild flavor, palatability, and mucilaginous quality of purslane give it a wide range of uses in the kitchen: as a vegetable, like spinach, and in stews and soups as a thickener. It can also be pickled with vinegar, spices, and sugar.

Purslane has a higher iron content than any other vegetable and is reputed to alleviate scurvy. It is widely distributed in gardens and in waste places. The name is a corruption of *portulaca,* from the Old French *porcelaine.*

Pussytoes *Antennaria plantaginifolia*

The feathery appendage of each seed resembles a miniature butterfly antenna, hence the Latin generic name. The Latin species name means "plantain-like leaves." The common name aptly describes the young flower heads. These are soft and downy, like a cat's paw.

Pussytoes

Puttyroot *Aplectrum hyemale*

Aplectrum, Greek for "without a spur," calls attention to this orchid as one of the few that are spurless. *Hyemale,* a Greek word signifying "of the winter," indicates that the single leaf remains green through the winter. The bulb of this orchid, with its offset corn, is suggestive of a piece of putty. Often several bulbs are joined together and resemble chunks of putty even more closely.

Puttyroot

Pyracantha *Cotoneaster pyracantha*
See Cotoneaster for the derivation of the generic name.
The red berries and prominent spines of this orna-mental shrub provided the basis for the specific and common names, made up of the Greek words for "fire" and "spine."

Pyracantha

Pyrola; Shinleaf *Pyrola elliptica*
The Greek name for this genus signifies "like a pear tree," an allusion to the resemblance of pyrola leaves to those of the pear. The species name refers to the elliptical leaves.
Pyrola leaves at one time were an ingredient of a plaster used to treat skin sores, hence the alternate name.

Pyxie; Pyxie Moss *Pyxidanthera barbulata*
The generic name is derived from the Greek *pyxis*, "a small box," and *anthera*, "anthers." The anthers of this plant open like the lid of a small box to shed their pollen. The bunches of short, needle-like leaves are suggestive of "a cropped beard," the meaning of the Latin *barbulata*.
The common name is an easy abbreviation of the lengthy generic name. This trailing, moss-like plant, with stemless white or pink flowers, is found in sandy pine barrens.

Pyxie

Quince, Flowering *Cydonia oblonga*
The quince supposedly originated near the town of Cydon, on the northwest coast of Crete; hence the generic name. *Oblonga* is descriptive of the fruit's shape. The origin of the word *quince* is uncertain; one authority believes it to be a corruption of the generic name.
This shrub is grown for its scarlet flowers and green-yellow fruit. The Japanese quince, *C. japonica*, is cultivated for its large, showy flowers.

Radish to Rutabaga

Garden Radish

Radish, Garden *Raphanus sativus*
The rapid germination of radish seeds gave rise to the Greek generic name, which means "quickly appearing." *Sativus* is Latin for "sown in fields or garden." Radish is a corruption of the Latin *radix,* meaning "root," which come through the French *radis.*

A wild plant in the cooler parts of Asia is believed to be the ancestor of this root vegetable. Several unusual forms have been developed, such as the Madras radish, with tender soft pods that are eaten raw or pickled, and the edible rattail radish with foot-long pods.

Ragweed, Common *Ambrosia artemisiifolia*
Ambrosia with nectar was the food and drink of the gods in ancient mythology. This name later was applied for unknown reasons by Linnaeus to this rank weed.

The Latin species name alludes to the artemisia-like foliage of ragweed. The ragged-appearing, deeply-incised leaves suggested the common name.

Ragweed leaves are reputed to have astringent detergent properties. The plant is best known and most despised as a cause of hay fever in late summer.

Ragweed, Giant; Buffalo-weed *A. trifida*
See Ragweed, Common, for the derivation of the generic and common names.

Trifida, meaning "three-parted," refers to the three-lobed leaves of this species. This widespread weed, a migrant from Europe, also is a bane of hay fever sufferers.

Common Ragweed

218

Rape, Garden *Brassica napus*
See Broccoli for the derivation of the generic name.
The species name was an old generic name for rape, which in turn derived from the Greek, *rhapys* and Latin *rapa,* meaning "turnip." Rape is grown as a forage crop and ground cover; and for rape oil and bird seed. It is widely naturalized.

Raspberry, Black *Rubus occidentalis*
See Raspberry, Purple-flowered, for the derivation of the common name.
The species name, meaning "western," distinguishes this as an American species. It is the parent of a number of cultivated varieties. The fruit is edible and the dried leaves make a refreshing tea.

Black Raspberry

Raspberry, Purple-flowered *R. odoratus*
See Blackberry for the derivation of the generic name.
The fragrant flowers account for the Latin species name. The rasp-like prickles on the stems and leaf-veins of several species explain the common name. This species has sticky, hairy stems and an edible fruit, inferior in quality to most other raspberries.

Purple-flowered Raspberry

Raspberry, Red *R. strigosus*
See Raspberry, Purple-flowered, for the derivation of the common name.
Strigosus, Latin for "bristly," distinguishes this from related species. It too is a parent of some of our cultivated varieties. It is used like black Raspberries as food and drink.

Red Raspberry

Rattlebox *Crotalaria sagittalis*
The ripe seeds of this legume rattle in the inflated hard pod when blown by the wind or disturbed by a passerby. This explains both the common name and the generic name; the Greek *crotalon* means "rattle." The arrow-shaped stipules at the leaf bases account for the Latin species name. Roasted rattlebox seeds

Rattlebox

are an acceptable substitute for coffee.

Rattlesnake Master; Button Snakeroot *Eryngium yuccifolium*
Eryngium, originally the Greek name for a thistle-like plant, later was applied to this genus. Its stiff and prickly yucca-like leaves explain the species name.

In early American folklore, an infusion made from this plant's roots was a reputed cure for snake-bite. The button-like flower head and its curative virtues account for the second common name.

C. W. F. Millspaugh, in his *American Medicinal Plants* (1887), reports that "a tea of boiled root is useful in remedying sexual depletion with loss of erectile power." Another author lists button snake-root as a diuretic.

Rattlesnake Plantain *Goodyera pubescens*
This native orchid was named in honor of John Goodyer, a seventeenth-century botanist who aided Johnson in the preparation of a new edition of Gerard's *Herball.* The Latin specific name alludes to the downy or woolly stem. The white reticulations on the leaves suggest the scale pattern of the rattle-snake.

Rattlesnake Root; White Lettuce *Prenanthes alba*
See Lion's-foot for the derivation of the generic name.

Alba, Latin for "white," describes the white drooping flowerheads. This plant was once believed to be distasteful to rattlesnakes or effective against their venom. The alternate common name reflects this plant's resemblance to wild lettuce.

Rattlesnake Weed; Hawkweed *Hieracium venosum*
See Devil's Paintbrush for the derivation of the generic name.

Venosum, Latin for "veined," alludes to the marked purple-veined leaves. The common name is explained by this quotation from an 1861 almanac:

Rattlesnake Plantain

"This weed is plentiful where the deadly reptile abounds."

Red-hot Cattail. See Chenille-plant

Red-hot Poker; Torch Lily *Kniphofia caulescens*
This red-flowered South African lily was named in honor of J. H. Kniphof (1704–1765), professor of medicine at Erfurt University in Germany. He was the author of a beautifully illustrated volume on plants in 1747. He also served as librarian of the Academy of Natural Curiosities. *Caulescens,* from the Latin *caulis* meaning "stalk," refers to the long stem of· this lily. The two common names allude to its bright red color.

Redroot. See Lamb's Quarters

Rhododendron, Catawba *Rhododendron catawbiense*
See Azalea for the derivation of the generic name. The common and species name are from the Choctaw Indian *katapa*. This Sioux group once inhabited an area along the Catawba River in South Carolina. This species is native of the mountains from Virginia to Georgia. The related *R. canadense* is found in swampy areas from Canada through New England. Over 4,000 varieties and species of rhododendron have been catalogued for North America. A closely related species, *R. maximum,* is the state flower of West Virginia. Another related species, *R. californicum,* is the state flower of Washington.

Catawba Rhododendron

Rhubarb, Garden *Rheum rhaponticum*
The full scientific name from Greek means "rhubarb of Pontus." The latter was a Roman province in Asia Minor near the Black Sea. The common name is a corruption of the old Greek name, stemming directly from Middle Latin *rheubarbarum.* Both *rha* and *rheon* are different forms of the same Greek word.
Rhubarb can be an ornamental as well as edible

plant. The leaf stems are used for pies, sauces, and desserts. The juice makes an excellent wine, a fact little-known in the United States.

Richweed. See Horsebalm

Robin's Plantain *Erigeron pulchellus*
See Fleabane, Daisy, for the derivation of the generic name.
Pulchellus, Latin for "beautiful," aptly describes this spring flower. The origin of the common name is uncertain. Jonathan Carver, in his *Travels* (1778), mentions a poor robin's plantain which "received its name from its size and the poor land on which it grew."

Rock Cress, Purple *Aubrietia deltoides*
This violet and purple rock garden plant commemorates Claude Aubriet (1668–1743), a French botanical artist and plant collector. In 1700 he went on an Asiatic plant-collecting expedition sponsored by Louis XIV "to discover the plants of the ancients, and others which escaped their knowledge."
He visited 33 Greek islands in search of plants and then went to Constantinople for outfitting. He joined a caravan bound for Erzurum and Trebizond (present-day Turkey), collecting specimens and making drawings of plants wherever he went. He was the first to describe and paint the azalea, rhododendron, and Oriental poppy. The last he introduced into Europe. The expedition visited Tiflis and Erivan, but failed in an effort to ascend Mt. Ararat. The party turned homeward, with 1,356 dried plant specimen, hundreds of drawings, and many seeds.
The specific Latin name refers to the triangular-shaped leaves. The common name arose from its resemblance to a cress and its preference for a rocky habitat.

Purple Rock Cress

Rock Cress, Smooth *Arabis laevigata*
Arabia was believed to be the original home of

the rock cress, hence the generic name. *Laevigata,*
Latin for "smooth," alludes to the leaf. The common name indicates an affinity for rocky terrain.
This weedy cress produces long, narrow curved
pods. The cultivated forms, popular in rock gardens,
bear white, crimson, and purple flowers.

Rocket, Yellow *Barbarea vulgaris*
This cress-like plant was once known popularly as
the Herb of Santa Barbara, the saint to whom it is
dedicated. It grows vigorously during short spells of
warm weather in winter and is often found growing
on this saint's day in early December, when almost
all other herbs are dormant. The rosettes of leaves
can be used in a salad or as a potherb. *Vulgaris,*
Latin for "common," alludes to the widespread occurrence of this species.

Rock Jasmine *Androsace carnea*
Androsace signifies "man" and "buckle" in Greek,
a reference to the anthers which supposedly resemble a buckle. The pink flowers inspired the species
name, Latin for "flesh-colored." This tufted, rock-
garden herb was fancied to resemble a jasmine,
hence the common name.

Rocky Mountain Garland *Clarkia elegans*
This genus honors William Clark (1770–1838),
who was born in Virginia but grew up on the Kentucky frontier. There he learned surveying, mapping,
and hunting and gained a knowledge of fauna and
flora. He enlisted in the army in 1792 and served on
several expeditions to punish the Indians for depre-
dations in the Indiana and Illinois territories. It was
during this service that he met Meriwether Lewis.

In 1803, Lewis invited Clark to join him on an
exploratory mission to find a route to the Pacific.
The expedition, which lasted two years, was an important event in the westward expansion of the
United States.

Clark later was named Superintendent of Indian

Rocky Mountain Garland

Affairs and in 1813 became governor of the Missouri Territory. He also served as an officer in the War of 1812.

Elegans, Latin for "elegant," aptly describes this mountain flower. The common name alludes to the resemblance of *Clarkia* to a garland of almond flowers. This species is used for mass plantings as well as for cut flowers.

Rose, Garden *Rosa spp.*

The name derives from the Latin *rosa* and the Greek *rhodon*. In ancient mythology the rose was the symbol of love and beauty. Chloris, goddess of flowers, crowned the rose as queen of flowers. Aphrodite presented a rose to her son Eros, god of love. Thus the red rose became a symbol of love and desire. Eros gave the rose to Harpocrates, god of silence, to induce him to hide the weaknesses of Greek gods. Thus the rose also became the emblem of silence and secrecy. In the Middle Ages a rose was suspended from the ceiling of a council chamber, pledging all present to secrecy, or *sub rosa,* "under the rose."

The rose became an important heraldic symbol. During the "War of the Roses," the House of York was symbolized by a white rose, the House of Lancaster by a red rose.

Japanese Rose

Rose, Japanese *Rosa rugosa*

See Rose, Garden, for the derivation of the generic and common names.

The specific name, from Latin, aptly describes the wrinkled leaves of this species, widely planted as a hedge or roadside shrub. The pulpy fruit, or hip, is used in jelly-making and is very high in vitamin C content.

Rose, Multiflora *R. multiflora*

See Rose, Garden, for the derivation of the generic and common names.

The species name, Latin for "many flowered," refers to the large number of flowers per stem. This Asiatic introduction, widely used as an ornamental, often escapes from cultivation.

Rose, Pasture or Carolina *R. carolina*

See Rose, Garden, for the derivation of the generic and common names.

This wild rose, first described from a southern specimen, is the state flower of Iowa, North Dakota, Georgia, and New York.

Pasture rose petals are used in making rose petal jam and candied rose petals. To make jam, combine petals, some water, lemon juice, and sugar or honey and mix in blender.

Pasture Rose

Rose Mallow, Swamp *Hibiscus palustris*

See Okra for the derivation of the generic name. *Palustris,* Latin for "swamp-dwelling," refers to the marshy habitat of this species. It produces hibiscus-like pink flowers up to eight inches in diameter. Some mallows are known locally as shoe-black plants because the petals are used to put a mirror polish on shoes.

Rosemary *Rosemarinus officinalis*

The generic name and the common name are based on two Latin words, *ros* and *marinus,* signifying "dew of the sea," an allusion to the native habitat of rosemary on the sea cliffs of southern Europe.

The species name refers to its former availability in the market stall and apothecary shop. This pungent, bitter herb was used as a flavoring for fish and sauces. In the Middle Ages a hot drink made of hot milk, ale, honey, and rosemary leaves was recommended as a heart stimulant and nerve tonic. Hot rosemary tea was prescribed to relieve headache. It also was the source of an early mouthwash. Gerard, in his *Herball,* advises that "The distilled water of the floures of Rosemary, being drunk at morning and evening first and last, taketh away the stench of the mouth and breath."

Rose Mallow

Rose Moss. See Portulaca

Rose-of-the-rockery *Geum reptans*
See Avens, Purple, for the derivation of the generic name.
Reptans, Latin for "creeping," refers to the prostrate growth habit of this yellow-flowered border and rock garden plant.

Rose Pink. See Marsh Pink

Roseroot *Sedum rosea*
See Goldmoss for the derivation of the generic name.
Rosea refers to the color of the root of this stonecrop, as does the common name. The succulent leaves are edible as a salad or potherb. Leaves and flower parts are in fours, and the flowers are purplish.

Rose of Sharon *Hibiscus syriacus*
See Okra for the derivation of the generic name.
This relative of hibiscus and mallow was erroneously believed to have originated in Syria, hence the species name. It was named for the Biblical rose of Sharon, to which it is unrelated.

Rose of Sharon

Rosinweed; Compass-plant *Silphium integrifolium*
See Indian Cup for the derivation of the generic name.
The Latin species name means "entire leaved," that is, the leaves are not lobed or divided. The common name relates to the resinous foliage and odor of this plant. The alternate name alludes to the inclination of this plant to present its leaf edges to the north and south.

Royal Basil. See Sweet Basil

Rubber Plant *Ficus elastica*

See *Fig* for the derivation of the generic name. *Elastica* refers to the latex-bearing sap from which rubber can be derived. This decorative house plant is very sturdy and enduring.

Rue Anemone *Anemonella thalictroides*

This generic name is the diminutive of *Anemone* or wind-flower, which this plant resembles. It also bears some resemblance to the meadow-rue, *Thalictrum;* hence the specific name. This member of the buttercup family produces delicate white flowers, and can be easily mistaken for a true anemone.

Ruellia *Ruellia strepens*

Jean de la Ruelle (1474–1537), a physician and botanist, was herbalist to François I of France and a canon of Notre Dame. He wrote *Of the Natural Races* and compiled several Greek and Latin works on botany. *Strepens,* Latin for "rattling or creaking," was bestowed as a specific name in apparent reference to the rattling of seeds in the dry pod.

Rue Anemone

Russian Thistle; Prickly Saltwort *Salsola kali*

This weed often is found growing in salty soil, hence the Latin generic name. The Arabic word *kali* means "alkaline," another habitat reference. This prickly, pesky plant, thought to have originated in northern Europe, is not a thistle and probably not of specific Russian origin.

Rutabaga *Brassica oleracea var. napobrassica*

See Broccoli for the derivation of the generic name and Cabbage for the species name.

The varietal name signifies "rape cabbage." The common name is related to the Swedish *rotabagge* and the French *rutabaga*. This hardy, yellow-fleshed root tuber was originally called *B. campestris* by Linnaeus.

Sage to Syngonium

Saffron, Meadow. See Crocus, Autumn

Sage, Garden *Salvia officinalis*
    Salvia is Latin for "safe" or "healthy," an allusion to the sage's long use as a medicinal herb. The species name refers to its place in the apothecary shop. The common name reflects the slow evolution of the original Latin name through *salvie, sauge,* and *salge.*
    This important herb of antiquity was dedicated to the Greek god Zeus and the Roman god Jupiter. At various times and places sage was prescribed for ailments of the blood, brain, liver, stomach, and heart, as a cure for epilepsy and fevers, and as a protection against plague.
    Its virtues have been deflated in modern times. Sage now has a place as a culinary herb, and sage tea is often used as a substitute for or change from India tea. Cultivated varieties are widely used in garden borders.

Sage, Lyre-leaf *S. lyrata*
    See Sage, Garden, for the derivation of the generic and common names.
    This wild species has lyre-shaped leaves. It is pungent and aromatic and also has use as a food flavor.

Sainfoin. See Scurf Pea

Saint Andrew's Cross *Ascyrum hypericoides*
    The Greek *a skyros,* meaning "not rough," is the

Saint Andrew's Cross

root of this generic name. The species name signifies that the leaves "resemble (those of) St. John's-
wort." The flower petals, in the form of a St. Andrew's Cross, explain the common name. For those
familiar with this wildflower, it should be noted that
most species are not soft to the touch; they are
roughish or hairy, contrary to the generic name.

Saint-John's-wort *Hypericum perforatum*
"Beneath or among the heather" is the translation
of this Greek generic name. The Saint-John's-wort is
often found growing near members of the heath family. The species name, meaning "having small holes"
(akin to the English word "perforated"), actually
refers to black dots on the petal edges and to the
translucent dots on the leaves. The species describer
apparently had deficient eyesight.

The common name arose through association of
the flowering period in June with the observance of
St. John's Day on June 24th.

This plant long was believed to have the power to
ward off witches and other evil spirits. Thus, in Norway and Sweden, Saint-John's-worts were gathered
and hung in the parlor during June, when nights
were shortest and witches held nocturnal festivities.

A coloring matter, extracted from the St. John's-
wort flowers, formerly was used in the treatment of
wounds.

Saint-John's-wort

Saint-John's-wort, Marsh *Triadenum virginicum*
The generic name, meaning "three glands," pertains to several sets of three orange glands in the
flower. The function of these glands is unknown.
They alternate with bundles of three stamens. *Virginicum,* signifies "of Virginia," the state from which
the species was described.

This close relative of the Saint-John's-wort is found
only in bogs and swamps from Labrador to Florida.

Saint-Peter's-wort *Ascyrum stans*
See Saint Andrews-cross for the derivation of the
generic name.

Saint-Peter's-wort

229

The specific name, signifying "standing or erect," is descriptive of the plant's posture. This species received its name because it flowers about the time of Saint Peter's feast day, June 29th.

Salmonberry *Rubus spectabilis*
See Blackberry for the derivation of the generic name.
The specific name, meaning "remarkable" or "notable," is descriptive of the large, often solitary flowers. The common name refers to the salmon color of the edible berry.

Salsify. See Oyster Plant
Saltbush. See Orache

Samphire *Salicornia europaea*
See Glasswort for the derivation of the generic name.
The species name denotes the European origin of this denizen of our coastal salt marshes. The common name is the contraction of the French *herbe de St. Pierre*. Samphire is used as a salad plant and may be pickled.

Sand Spurrey *Spergularia rubra*
The Latin word *spergula,* meaning "scattering of seeds," accounts for both the common and generic names. It refers to the manner in which the tiny seeds are scattered when the capsule opens. *Rubra,* Latin for "red," alludes to the sand spurrey's pink petals.

Sand Verbena *Abronia umbellata*
The Greek word *abros,* meaning "delicate," is descriptive of the bracts beneath the flowers. The species name tells us of the umbrella-like arrangements of the pink fragrant flowers. This verbena-like herb prefers a sandy habitat, hence the common name.

Sandwort, Rock *Arenaria stricta*

This plant, as the common name suggests, prefers a sandy habitat. Its generic name is based on the Greek word for sand, "arena." *Stricta* describes the upright growth habit of this species. Sandwort can be used as a salad plant, a potherb, and as a pickle. An allied species, *A. serpyllifolia,* bearing thyme-like leaves (the meaning of the specific name), also is edible.

Rock Sandwort

Sanicle. See Snakeroot, Black

Sarsaparilla, Wild *Aralia nudicaulis*

See Hercules'-club for the derivation of the generic name.

Nudicaulis, Latin for "naked stem," is descriptive of the lower part of the plant, which is devoid of leaves. The common name comes to us from the Spanish *sarza,* "a bramble," and *parilla,* "a little vine." Sarsaparilla is a climbing vine. Its berries have a slight resemblance to grapes. The rootstock is used in the home preparation of root beer and was once used as a stimulant and diuretic.

Satinflower *Godetia grandiflora*

C. H. Godet (1797–1879) was a noted Swiss botanist. In the 1850s he published *Flora of the Juras-Swiss and French,* and *Description of Plants of Neuchatel. Grandiflora,* meaning "large-flowered," aptly describes the large bright flowers of this garden annual. The common name also reflects the attractiveness of the blooms.

Wild Sarsaparilla

Savory, Summer; Basil *Satureja (or Satureia) hortensis*

The Latin generic name stems from the Arabic word *sattar,* meaning a kind of mint. *Hortensis,* "of the garden," reflects the long cultivation of savory as a garden herb. The common name stems from the Latin, through a process of word evolution.

The aromatic leaves of this herb are used in cook-

Summer Savory

ing and as a salad green. Powdered savory once
served as a flea repellant. Virgil, in his writings,
recommended it for planting near bee hives. Savory is
reputed to be a strong stomach tonic when steeped
in wine.

The alternate name, basil, Greek for "royal," has
an interesting history. This was looked upon as a
royal herb because it was used in the royal bath and
in regal medicine.

Savory, Winter *S. montana*

See Savory, Summer, for the derivation of the
generic and common names.

Montana, "of the mountains," refers to the alpine
habitat of this close relative of summer savory. It is
called winter savory because of its hardiness and
longer season of growth.

Saxifrage *Saxifraga virginiensis*

"I break rocks," the translation of this generic
name, refers to the supposed rock-cracking ability
of this denizen of rocky crevices. Another account
bases the name on its reputed use as a cure for gall-
stones. The specific name alludes to its best known
habitat, Virginia.

The young, unrolling leaves are tender and edible
as a salad.

Scabiosa *Knautia arvensis*

This genus was named in honor of Christian
Knaut (1654–1716), a physician and botanist in
Saxony. The species name denotes its habitat as
"cultivated fields."

The name *scabiosa* was conferred upon this plant
by Linnaeus, from a Latin word meaning "scratch,
itch, or scrape." This was a reputed remedy for
scabies and itch caused by the mange mite.

Saxifrage

Scorpionweed *Phacelia dubia*

The Greek word *phakelos,* meaning "a bundle,"
was chosen as the root of this generic name because

the flowers seemed to be grouped in bundles. The species name, meaning "dubious" or "doubtful," was assigned because it was looked upon as atypical of the genus.

The flower stalk is coiled or scorpion-shaped at the tip and uncoils as the flowers change into seeds. This phenomenon gave rise to the common name.

Screw Pine *Pandanus pygmaeus*

This generic name is a Latinization of the Malayan vernacular name, *pandan.* As the smallest species in the genus, this one received the Latin name for pygmy. This tropical shrub bears spirally arranged, sword-shaped leaves, hence its name. This species, not a true pine, grows only two feet tall and is a well-known houseplant.

The long use of pandanus in basket-making gave rise to the specific name of another species, *P. utilis,* meaning "useful." Pandanus has variegated leaves, marked with yellow or white.

Scurf Pea; Sainfoin *Psoralea onobrychis*

The botanist who bestowed the generic name on the scurf pea took note of the scurfy appearance of the leaves that is caused by numerous small black glandular dots, hence the Greek name meaning "scurfy." He also observed that grazing asses sought out this wild pea-like plant, so he devised a Greek species name signifying "ass bray," a supposition that asses brayed in joy in the presence of this fodder.

The alternate common name is of French origin, from *sainct-foin.* A quotation from Surflet (1600) indicates the origin of this name: ". . . for that it may seeme to spring out of the earth, and as it were, of a more speciall favor from God."

Sea-blite *Suaeda maritima*

An Arabic vernacular name, *suwayd,* is the source of this generic name. The species name denotes the sea-blite's salt marsh and sea beach habitat. The common name is based on an old Greek word, *bli-*

ton, which sea-blite supposedly resembled. This prostrate, fleshy leaved plant is edible in the spring as a potherb. The leaves and branches must be boiled twice to remove the excess saltiness.

Sea Holly *Eryngium planum*
See Rattlesnake Master for the derivation of the generic name.

Planum, meaning "flat," refers to the flat, yucca-like leaves of this seashore plant. The common name is descriptive of the spiny-edged leaves and the coastal habitat. The sweet, aromatic roots were once regarded as possessing aphrodisiac properties. During the sixteenth to eighteenth centuries the roots were used for candying. Gerard, in his *Herball,* stated that the roots, preserved with sugar, "are exceedingly good to be given to old and aged people . . . restoring the aged, and amending defects of nature in the younger." The sea holly also was believed to cure the bite of a rattlesnake, hence the name rattlesnake master in colonial America.

Sea Lavender

Sea Lavender; Statice *Limonium nashii*
The generic name of sea lavender is derived from the Greek *leimon,* "meadow," a reference to the salt marsh habitat of this species. George V. Nash (1864–1921), the honoree for the specific name, was a noted grass specialist who traveled widely in search of new grasses. He became curator of the New York Botanic Garden in 1899. The common name apparently relates to a fancied resemblance of this species to the lavender.

Statice, the alternate name, is usually applied to a cultivated form of *Limonium.* It is a Greek word meaning, "I stop." The astringent juice was at one time used in medicine, especially to treat dysentery.

Sea-milkwort *Glaux maritima*
Glaux, the ancient name of milk vetch, later was applied to this small herb of the salt marshes. *Maritima,* "of the seashore," aptly describes the habitat of sea-milkwort. The latter name is explained by a quotation from Henry Lyte's *A Niewe Herball*

(1578): "This, taken with meate, drinke or potage, ingendreth plenty of milk; therefore it is good to be used of nurses that lacke milke."

Sea Poppy. See Horn Poppy

Sea Purslane *Sesuvium maritimum*
   The name of an ancient Gallic tribe, for reasons not now clear, was used as a generic name for this seashore (*maritimum*) species. The sea purslane looks like the edible purslane but is not related. This prostrate weed forms a mat in damp, sandy places along the Atlantic coast.

Sea Rocket *Cakile edentula*
   Cakile is the Arabic vernacular name, applied as a generic name. *Edentula,* meaning "toothless," refers to the smooth edges of the leaves, that is, without teeth. Sea rocket received its common name because of its seaside habitat and some resemblance to the dame's rocket. The young foliage can be used as a salad, and the succulent leaves and branches can be cooked as a potherb.

Sea-milkwort

Seedbox; False Loosestrife *Ludwigia alternifolia*
   This genus commemorates a prolific researcher and writer, C. E. Ludwig of Leipzig University (1709–1773). He served as botanist on a voyage of exploration to Africa in 1731. He wrote *Observations on Plant Sexuality* in 1739. This was followed by fifteen other botanical works over the next two decades. Ludwig was named professor of medicine in 1740, and he later taught anatomy, therapeutics and physiology—a versatile teacher.
   The species name indicates an alternate arrangement of leaves on the stem. The common name refers to a resemblance to the true loosestrife. The name seedbox is descriptive of the short, squarish, loosely seeded capsule.

Seedbox

Self-heal. See Heal-all

Seneca Snakeroot

Wild Senna

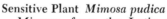

Sensitive Plant

Seneca Snakeroot *Polygala senega*
See Milkwort for the derivation of the generic name.

The specific name pays tribute to the New York Indian tribe from whom the white settlers first learned of the medicinal uses of this and other plants. Snakeroot was used by the Seneca Indians to treat snakebite poisoning. They chewed the roots and applied them as a wet mash to the bite. This herb was listed in the U.S. Pharmacopoeia for 116 years.

Women of the Ottawa and Chippewa tribes drank snakeroot tea to induce abortion in unwanted pregnancies. No reports are extant as to its effectiveness.

Senna, Wild *Cassia hebecarpa*
See Partridge Pea for the derivation of the generic name.

Hebecarpa, meaning "down-covered seed," is descriptive of the senna seedpod. *Senna* was borrowed from *sana,* the Arabic name of this plant. The laxative properties of senna were well known to the Indians.

Sensitive Plant *Mimosa pudica*
Mimosa, from the Latin *mimus* and the Greek *mimos,* meaning "to act," refers to the ability of the leaves to fold and the leafstalk to droop upon touch or other contact. *Pudica,* meaning "bashful" or "modest," also refers to the sensitivity alluded to above. The leaves and stalks regain their original posture in about an hour.

Sensitive Plant, Wild *Cassia nictitans*
See Partridge Pea for the derivation of the generic name.

Nictitans, meaning "blinking" or "moving," refers to this species' sensitivity to touch. The leaves fold tightly together when touched.

Sesame *Sesamum indicum*
The original Greek name for this medicinal herb

was continued as a generic name and in a slightly modified form as the common name. The species name, "from India," denotes the country of origin.

Sesame oil is widely used in cookery. The seeds are used in cookery and baking. The oil once was used in cosmetics.

For many centuries, in Asian countries, the sesame plant, soaked in sparrow eggs and boiled in milk, was believed to be an effective aphrodisiac.

Shallots *Allium ascalonicum*

See False Garlic for the derivation of the generic name.

The species name is the Latinization of Ashkalon, an ancient seacoast town now in Israel, believed to be the original home of shallots. The common name stems from Latin *ascalonia,* the Old French *eschaloigne,* and the Middle French *eschalotte.*

Shallots are prized in cookery and flavoring; the leaves are used in salads. The bulbs have a mild flavor and keep for a year.

Shamrock

See Trifolium and Oxalis.

Both white clover (*Trifolium repens*) and wood sorrel (*Oxalis europaea*) have been used by the Irish for many centuries as a symbol of Saint Patrick's Day. Both have three leaflets, which symbolically represent the Trinity. The clover is believed to have the edge in popularity today.

Shamrock is derived from old Irish *seamair,* meaning "clover," and its diminutive, *seamrog,* "little clover." This would give the clover the edge on being the true shamrock of Ireland.

Shepherd's Purse *Capsella bursa-pastoris*

All three names tell us something of the seed capsule of this plant. *Capsella* means "little box." The species name is Latin for "shepherd's purse" because the flat seed vessels resemble a pouch or purse. The English name is thus a straight translation of the scientific name.

The peppery pods add flavor to soups and salads. The young leaves, when boiled, are palatable and taste like cabbage.

Shinleaf. See Pyrola

Shooting Star *Dodecatheon meadia*
An ancient Greek plant name meaning "flower of the twelve gods" was bestowed on this attractive plant. The specific name honors Richard Mead (1673–1754), an English physician interested in botany. The striking reflexed corolla suggested the common name.

Shrimp Plant *Beloperone guttata*
This generic name, translated "arrow buckle" refers to the arrow-shaped connective between the stamen filament and the anther lobes, a connection that only a botanically trained reader is likely to understand. The species name, meaning "spotted" or "speckled," alludes to the flowers. The orange-coppery bracts, which have a marked resemblance to shrimp, gave rise to the common name.

Shooting Star

Sickle-pod *Arabis canadensis*
See Rock Cress for the derivation of the generic name.
The species name, "of Canada" indicates the locale of the first specimen to be described. Sickle-pod is widespread through North America and is edible in the spring as a salad or potherb.

Silverrod *Solidago bicolor*
See Goldenrod for the derivation of the generic name.
This species has flowers of two colors, some white, some yellow, hence the "bicolor."

Sickle-pod

Silverweed *Potentilla anserina*
See Cinquefoil for the derivation of the generic name.
Anserina, meaning "pertaining to geese," is based on the belief that geese relish this plant. This is borne out by two common names for silverweed, goose tansy and goose grass.
The general silvery appearance of this plant, especially the silvery-white under surfaces of the leaves, gave rise to the common name. Several American Indian tribes used the thick root as food. When cooked, it tastes like parsnip.

Silverweed

Singapore Holly *Malpighia coccigera*
This genus was named in honor of Marcello Malpighi (1628–1693), a distinguished naturalist of Bologna, Italy, who wrote on the anatomy of plants. The species name, meaning "berry-bearing," alludes to the round red berry. The holly portion of the common name refers to the spiny mature leaves which have some resemblance to holly. The locale is wrong, since the original habitat of this species was the West Indies, not Singapore.

Skeletonweed *Chondrilla juncea*
This is an ancient Greek plant name, derived from the word for "lump," an allusion to an exudate on the stems. *Juncea* means "rush-like." The flower head arises on a leafless, rush-like stem, from a basal rosette of leaves. Indian medicine men prescribed a tea made of dried leaves of this plant to encourage lactation by nursing mothers.

Skimmia, Japanese *Skimmia japonica*
Skimmi is the vernacular Japanese name of the first of this genus introduced to the Western world. This popular garden plant bears evergreen leaves and bright red berries. Male and female flowers are on separate plants.

Skirret *Sium sisarum*

See Parsnip, Water, for the derivation of the generic name.

The specific name is from *siser,* the Latin name for this herb. Emperor Tiberius commanded that this be brought to him from the Rhine every year, as he considered it a valuable medicinal plant.

The common name of this herb and garden vegetable traces to the Middle English *skirwhit,* which translates "pure white," and derives from the French *chervis,* the word for caraway. The sweet, white edible root is popular in Europe.

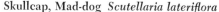

Skullcap, Mad-dog *Scutellaria lateriflora*

The Latin word *scutella,* meaning "small shield," aptly describes the fruiting calyx, or seed container. The specific name means "flowers on the side," an allusion to the arrangement of flowers on the stalk. The flower lover who popularized this fanciful common name saw a canine skullcap in the shape of the flower and made it more fanciful with the adjective mad-dog. This American mint was widely used at one time as a tonic and antispasmodic.

Mad-dog Skullcap

Skunk Cabbage *Symplocarpus foetidus*

The manner in which the ovaries grow together to make one knob-like fruiting stalk accounts for the generic name, from the Greek *symploke* and *karpos,* meaning "connected fruit." *Foetidus,* meaning "evil-smelling," is appropriate for the skunk cabbage. The odorous cabbage-like leaves gave rise to the common name. The leaves release the odor only when crushed ·or bruised.

Despite its name, this swamp herb is edible as a cooked vegetable. The leaves must be boiled in two or three waters, with a teaspoon of baking soda added, to dispel the pungent odor and flavor. Skunk cabbage was once prescribed as a remedy for respiratory disorders and rheumatism.

Skunk Cabbage

Skyrocket; Gilia *Gilia rubra*

This genus was named for Felipe S. Gil, an eigh-

Sage to Syngonium

teenth-century Spanish botanist. *Rubra* refers to its bright red flowers, which also is the basis for the common name. Several species, natives of California, are cultivated as border or edging plants.

Smartweed; Marsh-pepper *Polygonum hydropiper*
See Buckwheat, False, for the derivation of the generic name.
The specific name has been translated to the alternate common name. Smartweed received its name because it smarts or burns the tongue when tasted. For this reason the finely chopped leaves are used as a substitute seasoning.

Smartweed, Swamp *P. coccineum*
See Buckwheat, False, for the derivation of the generic name.
Coccineum, meaning "purple-flowered," aptly describes this swamp-dwelling species.

Snakemouth. See Pogonia, Rose

Snakeplant *Sansevieria trifasciata*
This genus was named in honor of Raimond de Sangro, Prince of Sanseviero, born in Naples in 1710, died 1771. The species designation, meaning "three-banded," refers to the banding on the leaf. The striped and barred leaves suggested the common name.
This decorative pot plant can endure neglect and poor light. If pampered, it may send up a spike of attractive white flowers in gratitude.

Swamp Smartweed

Snakeroot, Black; Sanicle *Sanicula marilandica*
The generic name, Latin for "healing," refers to the reputed medicinal properties of the sanicle. The species name, "of Maryland," records the locale of the first specimen described.
An anonymous colonial chronicler wrote of the snakeroot: ". . . an excellent preservative against poyson, called by the English the snake-roote." This

241

versatile root also was reputed to have astringent, antispasmodic, and antiperiodic properties.

White Snakeroot

Snakeroot, White *Eupatorium rugosum*
See Boneset for the derivation of the generic name.
Rugosum, meaning "wrinkled," refers to the somewhat curly leaves. This plant has been confused with other snakeroots, hence the common name.

Snapdragon *Antirrhinum orontium*
An imaginative botanist saw a resemblance of this flower to an animal's snout, thereby establishing the generic name. The slight resemblance of the leaves to those of *Orontium* (golden seal) accounts for the specific name.
This oddly shaped flower has borne several different names through the centuries and in different lands. The Romans fancied the flower as a lion's gaping mouth and named it lion's-snap. It also was known as calf-snout, dog's-mouth, and dragon-snap ("snap," meaning mouth or jaw).
At some point in history, a botanist with an ear for euphony transposed the name to snapdragon, a name which quickly gained ascendancy over others and has persisted to this day.

Snapweed. See Impatience

Sneezeweed, Bitter *Helenium amarum*
See also *Sneezeweed, Common.*
This is distinguished from the following species by its numerous narrow leaves. Its species name, meaning "bitter," refers to the fact that cows that eat this herb produce bitter, unmarketable milk.

Sneezeweed, Common *H. autumnale*
This genus was named in honor of Helen of Troy. The species name indicates the season of flowering. The odor of this tall, stout herb causes some passersby to sneeze, hence its common name.

Snowberry *Symphoricarpus albus*
See Coralberry for the derivation of the generic name.

The species name, meaning "white," and the common name refer to the milk-white berries.

In colonial days, a solution of crushed stems was applied to festering sores. A decoction of snowberry was also prescribed as a remedy for vomiting during pregnancy.

Snowdrop *Galanthus nivalis*
Galanthus, Greek for "milk flower," is truly descriptive of this early spring flower. Its specific name, meaning "from near the snow line," refers to its original alpine habitat in Europe and Asia. Snowdrop does well in cool, shady places.

Snow-on-the-mountain *Euphorbia marginata*
See Crown of Thorns for the derivation of the generic name.

The species name, meaning "margined" or "striped" alludes to the white-margined leaves. The colorful common name likewise is based on this leaf characteristic.

Snow-on-the-mountain

Soapwort. See Bouncing Bet
Solomon's Lily. See Black Calla

Solomon's Seal *Polygonatum biflorum*
The Greek words *poly* and *gonu,* meaning "many joints," allude to the numerous joints in the rootstock. *Biflorum,* "two-flowered," informs us that the flowers occur in pairs at the leaf bases. Some of the knobs and joints along the thick white rootstock have a resemblance to the mark of a seal. Thus, early Greek writers referred to this plant as Solomon's Seal.

Dioscorides stated that the roots were helpful in sealing or closing up fresh wounds, thus originating an alleged medical property widely accepted for many centuries. Gerard, in his *Herball,* believed that, "the root of Solomon's Seal, stamped [crushed] while

Solomon's Seal

243

it is fresh and green, and applied, taketh away in one night, or two at the most, any bruise, blacke or blew spots gotten by fals or women's wilfulnesse in stumbling upon their hasty husband's fists, or such like."

Earlier, the Greek physician Galen wrote that "if any of what sex or age soever chance to have any bones broken . . . their refuge is to stamp the roots hereof, and give it unto the patient in ale or drink; which sodereth and glues together the bones in very short space. . . ."

Both young shoots and roots have culinary use. Gather the shoots in early spring when they are about six inches long. These can be used in soups or stews or be served like asparagus with a dressing. Cut them up and boil them in salted water for 10 to 15 minutes. The fleshy roots must be boiled 20 to 30 minutes, after first cutting them up. They add an interesting flavor to a main dish.

Sorrel, Sheep *Rumex acetosella*

See Dock for the derivation of the generic name.

Acetosella, meaning "slightly acid," refers to the rather pleasant acidy taste of the leaves. The common name is a diminutive derived from the French word *sur,* meaning "sour." Thus we have the "little sour" (plant) relished by sheep.

Both wild and cultivated sheep sorrel have long enjoyed popularity as food plants. The wild plant is a pleasant nibble and thirst quencher. Sorrel or schav soup (the latter a Russian dish) is made of sorrel leaves, milk or water, grated onions, eggs, salt, and pepper. The cut up raw leaves are used as seasoning for a salad or fish platter.

Southernwood *Artemisia abrotanum*

See Dusty Miller for the derivation of the generic name.

Abrotanum is an old Latin plant name assigned as a species name. The common name reflects the origin of this genus in southern Europe. It is also known as smellingwood because of the aromatic, lemon-like odor of the foliage. Southernwood once was widely cultivated for medicinal use and as an ingredient in beer.

Sow Thistle *Sonchus oleraceus*

Sonchus is an ancient Greek plant name. *Oleraceus* means "garden or pot herb." The common name is from the Old English *sieger,* "sow," and *thistel,* "thistle," the latter alluding to the prickly leaf edges, the former to the sow's preference for this weed. The sow thistle is edible in early spring as a potherb, prepared like chickory, dandelion, or wild lettuce.

Soybean *Glycine soja*

The Greek word *glykys,* meaning "sweet," is the root of the generic name and alludes to the sweet-tasting root. The specific name and the first syllable of the common name are based on the Japanese vernacular name *shoyu* and the Chinese *shu.* The soybean has escaped from cultivation in those areas where it is grown as a crop.

Sow Thistle

Spanish Needles *Bidens bipinnata*

See Beggar-ticks for the derivation of the generic name.

The species name, Latin for "twice feathery," is descriptive of the double feathery arrangement of the leaves. The introduction of this noxious weed, which bears barbed seeds, was wrongly attributed to the Spaniards. Each seed has three to four curved barbs, helpful in obtaining their wide distribution by attachment to passersby and animals.

Spanish Needles

Spatterdock *Nuphar advenum*

Nuphar is from the Arabic vernacular name for this plant, *neufar. Advenum,* meaning "newly-arrived," refers to the immigrant status of spatterdock. The latter name is believed to indicate a dock-like plant that spatters its ripe seeds. These large seeds are edible if parched for ten minutes to loosen the kernels, then pounded lightly and winnowed to remove the shells. The starchy rootstock can be boiled, baked, or roasted as a vegetable. Remove the rind before eating.

Spatterdock, Small; Pond-lily *N. microphyllum*

See Spatterdock for the derivation of the generic and common names.

The species name, meaning "small-leaved," distinguishes this allied species. It also is edible.

Spearmint *Mentha spicata*

See Mint, Apple, for the derivation of the generic name.

The arrangement of the flowers in a tight spike explains the species name. According to Turner's *Herbal,* written in the sixteenth century "spearmint is so-called because of its sharp-pointed leaf."

This widespread native of Europe has many culinary and medicinal uses, among them mint jelly, aspic, tea, and candy flavor. Finely chopped mint is widely used in salads in the Middle East. It has been used to cure scurvy and dietary deficiency diseases because of its high vitamin A and C content.

Spearwort

Spearwort *Ranunculus ambigens*

See Buttercup for the derivation of the generic name.

The species name signifies that the sepals differ markedly from the petals. This mud-loving buttercup bears spear-shaped leaves, hence the common name.

Speedwell, Common *Veronica officinalis*

See Brooklime for the derivation of the generic name.

Officinalis means "of the market place or apothecary shop." Speedwell is a prostrate plant, and its name reflects its ability to spread speedily over the ground. A number of species arrived on our shores as stowaways in grain seed and ballast. Its use as a tea substitute is known, and it is reputed to possess diuretic and astringent properties.

246

Speedwell, Purslane *V. peregrina*
See Speedwell, Common, for the derivation of the generic and common names.

Peregrina, meaning "immigrant," refers to this species' original status. Widespread over the United States, it is a troublesome lawn pest, as it peregrinates over the grass. The purslane portion of the name denotes a resemblance of the speedwell leaves to those of the succulent purslane.

Spiderflower *Cleome spinosa*
Cleome is based upon the Greek *kleio,* "I shut," a reference to the flower parts; *spinosa* refers to the little spines in place of stipules at the leaf bases. The common name alludes to a notable characteristic of this flower—its long stamens, almost three times the length of the petals, possibly a record in flowerdom. Spiderflower frequently escapes from cultivation.

Purslane Speedwell

Spider Lily *Hymenocallis calathina*
This generic name, translated as "beautiful membrane," refers to the cup-like membrane which unites the stamen filaments—a somewhat technical detail. The flower was fancied to be "basket-shaped," the translation of the Greek *calathina.* The very narrow, almost linear flower parts account for the common name of this amaryllis relative, a native of Peru.

Spider Lily, Inland *H. occidentalis*
See Spider Lily for the derivation of the generic and common names.

Occidentalis, meaning "western," identifies a native species which occurs in marshes from Indiana southward. It bears an umbel of white flowers.

Spiderflower

Spider Plant *Chlorophyton elatum*
The generic name, Greek for "green plant," offers nothing distinctive about this decorative house plant. *Elatum,* meaning "tall," refers to the spider plant's erect growth habit. The common name derives from the fact that small plants develop at the end of long

247

wiry runners. These can be removed and planted. The narrow leaves often have light lines along the margin or colored bands down the center.

Spiderwort

Spiderwort *Tradescantia virginica*

John Tradescant was royal gardener to Charles I of England. He created the Museum Tradescantianum, a collection of rare and unusual plants, in his garden at Lambeth in 1656. He brought many exotic plants to England, including the apricot, gladiolus, and spiderwort. This generic name appears to honor two John Tradescants, father and son, both botanists and gardeners.

The son went on a collecting expedition to Virginia. He brought back the first specimen of spiderwort, hence the species name.

The common name alludes to the long twisting and jointed stems produced by this plant. The succulent stems and leaves can be used as a potherb.

Spikenard *Aralia racemosa*

See Hercules Club for origin of *Aralia*. The species name is descriptive of the flower arrangement, in loose clumps or racemes. Spikenard is from the Middle Latin *spica* and *nardi,* meaning "grain" and "ointment." The name traces back to a fragrant ointment of the ancients, mentioned in the Bible, which is attached in modern times to this aralia. The berries are used in making jelly. The aromatic root is a popular ingredient in homemade root beer.

Spinach *Spinacia oleracea*

The spiny seed clusters account for the generic name and, derivatively, for the common name. The species name identifies spinach as a garden or potherb. The Arabic word *isbanakh* suggests another possible derivation for spinach.

Spinach, New Zealand *Tetragonia expansa*

The well-defined four-angled seed is the root of the generic name; the specific name, *expansa,* alludes to its broad leaf.

This contribution from New Zealand provides spinach-like greens in summer when our spinach is no longer available. It has escaped from cultivation in many places in the South.

Spotted Knapweed. See Star Thistle

Spring Beauty *Claytonia virginica*

John Clayton (1686–1773), an early American botanist and physician, is recalled by this genus. Clayton came to Virginia in 1705 and was named county clerk of Gloucester County. He collected extensively in the colony; most of his specimens were sent to the National Herbarium in England.

Clayton's collection was used by Gronovius in preparing his *Flora Virginica,* the first scientific work on the subject. The delicate pink flowers, appearing early in spring, naturally suggested the common name.

Spurge, Cypress *Euphorbia cyparissius*

See Spurge, Leafy, for the derivation of the common name and Crown of Thorns for the generic.

The specific and common names are descriptive of the cypress-like foliage of this species. The minute flower clusters are surrounded by a pair of broad yellowish bracts.

Cypress Spurge

Spurge, Flowering *E. corollata*

See Spurge, Leafy, for the derivation of the common name and Crown of Thorns for the generic.

The specific name means "corolla-like." The showy flowers are really leafy bracts around minute flowers. These showy bracts account for the "flowering" portion of the name.

Spurge, Laurel *Daphne mezereum*

See Daphne for the derivation of the generic name.

The specific name is related to and possibly derived from the Arabic *mazariyum,* "camellia." Me-

Flowering Spurge

Leafy Spurge

zereum is also the name of a drug obtained from this plant. This small garden shrub, a native of Eurasia, bears fragrant purple-pink flowers, which are followed by red berries.

Spurge, Leafy *E. esula*

See Crown of Thorns for the derivation of the generic name.

The species name is based upon the French equivalent, *esule.* Spurge is derived from the French *epurge* and relates to the purgative properties for which this plant was known and used.

Corn Spurrey

Spurrey, Corn *Spergula arvensis*

The Latin *spergula* means "scattering of seeds," which this plant does well, to the farmer's dismay. *Arvensis,* "of cultivated fields," refers to the preferred habitat of this weed. Spurrey is a corruption of the generic name, while "corn" is the British term for "grain." In other words, this is a major pest of grainfields in Europe. Picturesque alternate names are devil's guts and farmer's ruin.

Spurrey, Sand *Spergularia rubra*

This generic name has the same derivation as corn spurrey, though it is not as noxious a pest. *Rubra* is descriptive of the pink flowers. This plant is common in sandy soil near the seacoast, hence the common name. It is grown as a soil renovator in Europe because it matures in twelve weeks. In Holland it is planted to stabilize shifting seashore sands.

Squash, Crookneck *Cucurbita moschata*

See Gourd for the derivation of the generic name.

The species name means "having a musky odor," a doubtful application here. We are indebted to the Algonquin Indians for the common name. Their word *askutasquash* was shortened to the last syllable by early Massachusetts Bay settlers.

Squash, Winter *C. maxima*
See also Squash, Crookneck.
Maxima, meaning "largest," indicates this as the weightiest of the squashes.

Squawroot *Conopholis americana*
The scaly, cone-like stalk of this plant accounts for the generic name, made up of two Greek words meaning "scaly cone." This leafless plant is parasitic on oak and hemlock roots. The common name reflects its use by the Indians as a remedy for various female disorders.

Squill *Scilla amoena*
The Greek root of this generic name means "I injure," a reference to the poisonous properties of squill. The species name signifies that these hyacinth-like flowers are "pleasing and lovely."

Squawroot

Squirrel Corn *Dicentra canadensis*
See Bleeding Heart for the derivation of the generic name.
The specific name, meaning "of Canada," refers to the locale of the first specimen described. The yellow tubers resemble those of Indian corn. There is doubt that squirrels eat these.

Starflower *Trientalis borealis*
The Latin *triens,* meaning "one-third," is an allusion to the low stature of this plant. *Borealis,* "of the north," indicates the habitat as the northern United States. This woodland beauty bears a six- to seven-pointed, white star flower atop a whorl of six to seven leaves.

Star Grass. See Colicroot

Starflower

Star Grass, Yellow *Hypoxis hirsuta*
Two Greek words meaning "sharp beneath," which is descriptive of the seed pod, make up the

generic name. The hairy (hirsute) leaves account for the specific name. The shape and color of the flower, a wild relative of the daffodil, provide the basis for the common name.

Star-of-Bethlehem *Ornithogalum umbellatum*
"Bird milk" is the literal meaning of the generic name, probably conferred whimsically in reference to the milk-white flowers. The arrangement of the flowers in umbrella-like fashion, that is, all arising from one point on the stalk, explains the specific name.

The common name was bestowed because of the supposed resemblance of the flower to the star, which according to St. Matthew, guided the wise men from the east to Bethlehem.

Star-of-Bethlehem

Star Thistle; Spotted Knapweed *Centaurea maculata*
See *Star* Thistle, Yellow, for the derivation of the common name.

The specific name, meaning "spotted," alludes to the bracts which are spotted at the tips and beneath the flower heads. This species is distinguished by its large pink to white flower heads. Knapweed traces back to the Anglo-Saxon word *cnaepp,* "knob" or "button," a reference to the knob-like flower heads.

Star Thistle, Yellow *C. solstitialis*
See Bachelor's Buttons for the derivation of the generic name.

The species name means "happening or appearing at the solstice" (June 22). This name is appropriate because the star thistle is at the height of its flowering in mid-June.

The common name is indicative of the thistly nature and yellow flowers of this weed. The ripe flower heads bear vicious spines which can injure people or livestock, especially in late fall or winter when they may be on the ground. This pest is commonly found in fields and pastures.

Star Tulip. See Mariposa lily

Statice. See Sea Lavender

Statice, Russian *Limonium latifolium*
See Sea Lavender for the derivation of the generic name.
Latifolium means "broad-leaved." This species has delicate lavender flowers which can be used as cut flowers or dried for winter decorative use. The common name indicates an ancestral Russian origin. The Greek word *statice* means "causing to stand." The finger-like spikes of flowers "stand" or remain fresh-looking for many months.

Steeplebush; Hardhack *Spiraea tomentosa*
See *Spiraea* for the derivation of the generic name. The species name, "densely woolly," describes the foliage. The terminal flower stalk resembles a steeple, hence the common name. The alternate name is of unknown origin.

Steeplebush

Stickseed; Beggar's-lice *Hackelia virginiana*
This genus was named in 1893 in honor of Paul Hackel, a Bohemian scientist. The species name indicates that it was first described from a Virginia specimen. The numerous small nutlets are armed with prickles, enabling the seeds to become attached to the clothing of passersby, hence the two common names.

Sticktights. See Beggar-ticks

Stitchwort *Stellaria graminea*
See Chickweed for the derivation of the generic name.
The specific name, meaning "grass-like," refers to the narrow leaves. The common name traces back to the Old English *stice,* a "stab" or a "stitch in the side," and *wyrt,* a "herb." This herb long was reputed to relieve discomfort from "stitches in the side" or other causes. A decoction of the root was used as a pain killer.

Stitchwort

253

Stock, Brompton *Matthiola incana*

Piero A. Mattioli (1500–1577), an Italian botanist and physician, wrote important works on medical botany and was a noted commentator on Dioscorides. *Incana,* signifying "hoary" or "gray," is descriptive of the appearance of the foliage of this cultivated ornamental with aromatic flowers. *Stock* is derived from the Anglo-Saxon *stoce,* meaning "stock, trunk, or stick." Brompton, a suburb of London, is the locale where this variety originated.

Stock, Night-scented *M. bicornis*

See Stock, Brompton, for the derivation of the generic and common names.

The species name, meaning "two-horned," refers to the pair of horn-like spurs on this variety of stock. The common name indicates that the aroma is evident at night, perhaps to attract nocturnal insects.

Stokesia *Stokesia laevis*

This genus honors Dr. Jonathan Stokes (1755–1831) of Chesterfield, England, a physician and botanist who successfully combined the two professions. In 1812 he published a four-volume *Botanical Materia Medica.* In 1787 Stokes prepared the *Botanical Arrangement of British Plants* jointly with Withering. A year before his death he had completed volume one of *Botanical Commentaries. Laevis,* "smooth," refers to the leaves of this popular garden perennial. *Stokesia* produces deep, blue flower heads that resemble asters.

Stonecress *Aethionema grandiflorum*

This generic name means "scorched filament" in Greek and refers to the appearance of the stamens. *Grandiflorum* suggests that the stonecress bears large flowers.

The common name alludes to its preference for stony or rocky locations and to its resemblance to cress.

Stonecrop. See Goldmoss

Stonecrop, Mountain *Sedum ternatum*
See Stonecrop, Stringy, for the derivation of the common name.
Ternatum, meaning "in threes," indicates that the fleshy leaves occur in whorls of three. This species bears a stalk of white flowers. This sedum, unlike other species, prefers moist rocky locales on mountain slopes, hence the common name.

Stonecrop, Stringy *S. sarmentosum*
See Goldmoss for the derivation of the generic name.
Sarmentosum means "bearing runners" and indicates the manner in which this species spreads. This rock-loving plant often covers rocks and walls, a characteristic which gave rise to its common name. This stonecrop, with yellow-green, starry flowers, frequently escapes from cultivation.

Mountain Stonecrop

Storksbill; Heronsbill *Erodium cicutarium*
This relative of the geranium and cranesbill also has a seed which looks like the beak and head of stork or heron. This fact explains the common names as well as the generic name, based upon *erodios,* Greek for "heron." The species name means "resembling the water hemlock." This is based on a similarity in the finely divided leaves. Cultivated forms of *Erodium* are used in borders and rock gardens.

Storksbill

Strawberry, Wild *Fragaria virginiana*
The Latin *fraga,* "strawberry," is based upon *fragrans,* "fragrant," a reference to the aroma of the fruit. The species name denotes the state from which the strawberry was first described. All cultivated forms stem from a hybrid of this species and the Chilean strawberry.

Strawberry Bush. See Burning Bush

Strawberry Geranium *Saxifraga sarmentosa*
See Saxifrage for derivation of generic name.

Wild Strawberry

The species name, Latin for "bearing runners," aptly describes the manner in which this plant spreads, which is much like the strawberry. The rounded, white-veined leaves closely resemble those of geranium. Many small white flowers are borne on one erect stalk, a characteristic which explains Aaron's-beard, an alternate common name.

Strawflower; Everlasting *Xeranthemum anuum*
The Greek generic name, translated "dry flower," is descriptive of this everlasting, one of the best-known of its class. The large scales on the flower head are petal-like and persistent, giving the plant its value as a dry decorative bouquet. There are white, rose, and purple varieties. *Anuum* indicates this is an annual.

Strawflower; Immortelle *Helichrysum bracteatum*
The attractiveness of these flowers is indicated by the Greek generic name, meaning "sun" and "gold." The bracts around the flower head are the basis for the species name, Latin for "with bracts." This popular garden and winter bouquet plant has large flowers in six different colors.

Sundew, Roundleaf *Drosera rotundifolia*
The Greek word *droseros,* meaning "dewy," alludes to the glistening dew-bedecked leaves with gland-tipped hairs. The appearance of the leaves on a bright, sunny day also explains the common name. The species name means "round-leaved."
The sundew is a bog-dwelling, insectivorous plant. A small insect, lured by the gland-tipped hairs, lands on the leaf in search of nectar. The longer marginal hairs bend inward, pinning down the luckless insect. The enzymous juices that exude from the leaf slowly absorb the nutritious innards of the insect.

Roundleaf Sundew

Sundew, Thread-leaved *D. filiformis*
See Sundew, Roundleaf, for the derivation of the generic and common names.

This species with thread-like leaves is found growing side-by-side with the roundleaf Sundew. It, too, captures small insects for sustenance.

Sundrops *Oenothera fruticosa*
See Primrose, Evening, for the derivation of the generic name.
The specific name signifies "bushy," a reference to the appearance of this plant. The attractive yellow flowers and orange stamens gave rise to the common name of this day-blooming relative of the evening primrose.

Sundrops

Sunflower, Common *Helianthus anuus*
The common name is a translation of the Greek generic name. The species name indicates that the sunflower is an annual. Pizarro, the conqueror of Peru, found the Incas venerating the giant sunflower as an image of their sun god. Priestesses, maidens of the sun, wore sunflower discs made of gold. The Spaniards eagerly sought these discs and took back seeds of this sunflower, the probable ancestor of the cultivated varieties. The edible seeds are a source of sunflower oil. A color-fast yellow dye is made from the ray flowers. The dried leaves are used as a tobacco substitute. In rural areas, the empty seed husks are used to make a coffee-like beverage. Sunflower seed sprouts are useful in cookery. This is the state flower of Kansas.

Sunflower, False. See Heliopsis

Sunflower, Mexican *Tithonia rotundifolia*
The generic name comes from Greek mythology. Tithonus was a young man beloved by Aurora, goddess of the dawn. This native of Mexico bears rounded leaves (hence the *rotundifolia*) and brilliant, orange-yellow flower heads, up to three inches in diameter.

Sun Rose *Helianthemum nummularia*

See Frostweed for the derivation of the generic name.

The specific name means "resembling money-wort," that is, the leaves are round and coin-shaped. The common name arose from the generic name and the fact that this species produces attractive yellow flowers in late summer. There are about 120 species of sun rose; those in cultivation used as ground cover, in rock gardens, and in borders.

Swallowwort, Black *Cynanchum nigrum*

This Greek generic name means "to poison or strangle dogs," a reference to the fact that this plant is poisonous to animals. *Nigrum,* meaning "black," is a misnomer for this species, since its flowers are purple-brown. The pods resemble a swallow with outspread wings, hence the common name. The German equivalent *schwalbenwurtz,* has the same derivation. This climbing vine is allied to the milk-weed because it bears silk-filled pods with tiny seeds; the silk serves as a parachute to distribute the seeds widely.

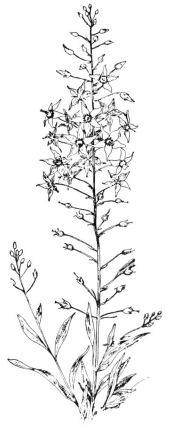

Swampcandle *Lysimachia terrestris*

See Moneywort for the derivation of the generic name.

Terrestris, meaning "terrestrial," is not quite appropriate, since this species prefers grassy shorelines and other wet places. The common name was suggested by the bright yellow flower spikes of this swamp-dwelling plant.

Sweet Basil; Royal Basil *Ocimum basilicum*

Ocimum is a classical Greek plant name which refers to the aromatic foliage. Basilicum, meaning "princely," alludes to basil's healing properties. The scientific names gave rise to the common names.

This aromatic annual has a long history of cultivation for culinary and medical use. It is used as a potherb and for seasoning of soups and stews. Basil oil is used as a flavoring agent and in perfumery. It

Swampcandle

was a sacred plant in Hindu gardens, where it played an important role in religious rites.

Sweetbrier *Rosa rubignosa*
See Rose for the derivation of the generic name. The specific name, meaning "reddish" or "rust-colored," describes the flowers. This fragrant European rose bears stout prickles, which account for the common name. It is naturalized along roadsides and in thickets.

Sweet Cicely. See Cicely, Sweet

Sweetbrier

Sweet Clover, White *Melilotus alba*
See Sweet Clover, Yellow, for the derivation of the generic and common names.
This species bears white flower spikes, hence the *alba* in its name. The dried flower heads are made up into sachets for use in the linen closet. The contraction *melilot* is also used as a common name.

Sweet Clover, Yellow *M. officinalis*
The fanciful Greek generic name means "honey lotus." This is partly in reference to the fact that bees are attracted to the flowers. The species name indicates its sale in the marketplace or apothecary shop. The growing shoots, gathered in spring before the flowers appear, are used in soups and as a pot-herb. In late summer, the ripe flower heads with tiny peas are used in bean or split-pea soup. The vanilla-flavored extract is used in baking cookies and pastries.

Yellow Sweet Clover

Sweet Coltsfoot; Butter-bur *Petasites palmatus*
The large leaves of this herb suggest a "broad-brimmed hat," hence the Greek *petasos* as the root of the generic name. The leaves also are involved in the species name, since *palmatus* means "palm-shaped." The country folk who devised the common name saw the leaf shape as that of a colt's foot, un-

aware of the discrepancy between this name and the scientific name. The foliage and emerging flower stalk are used as a potherb in spring and early summer.

Sweet Fennel. See Fennel, Florence

Sweet Flag; Calamus-root *Acorus calamus*
At one time calamus-root was prescribed as a remedy for eye ailments. This is the basis for the generic name, Greek for "pupil of the eye," preceded by the negative "a." The meaning apparently was that the root decoction would dispel ailments of the pupil of the eye. *Calamus* means "reed-like" leaves.

Candied flagroot was popular with rural folk in the nineteenth and early twentieth centuries. The Pennsylvania Germans use the root to flavor pickles. The inner parts of young shoots are used in salads.

The American Indians used the mashed, boiled rootstock in treating sores and burns, and the root extract was taken to induce abortion.

Sweet Gale *Myrica gale*
See Bayberry for the derivation of the generic name.

The species and common name derive from the German *gagel*, the name for sweet gale. The nutlets ripen in the fall and are used as a spice. The dried leaves make a good tea.

Sweet Flag

Sweet Olive *Osmanthus fragrans*
Osmanthus, Greek for "fragrant flower," is an apt designation, reinforced by the Latin species name and the "sweet" preceding "olive." This ornamental shrub related to the olive is grown for the handsome foliage and fragrant white flowers. It blooms the year-round.

Sweet Potato; Yam *Ipomoea batatas*
See Man of the Earth for generic name derivation.

The species name is the basis of the common name, in turn borrowed from the Spanish *batata,* a Carib Indian name for this vegetable. *Yam* is from the original Senegalese *inhame,* meaning "to eat." Slaves are thought to have brought this word into the English vocabulary.

The sweet potato is related to the morning glory, though it is not a climbing vine. It grows in the southern half of the United States and is harvested at first frost, when the leaves turn black.

Sweet Rocket. See Dame's Rocket

Sweetshrub; Carolina Allspice *Calycanthus floridus*

The closed calyx tube, suggestive of a cup, accounts for the generic name, Greek for "cup flower." *Floridus,* "flowery," informs us that this shrub bears showy flowers. The strawberry-scented leaves and twigs are dried and used in sachets. This shrub, native of the southeastern United States, is widely used in gardens and in landscaping in the South. Both common names reflect its aromatic qualities.

Sweetshrub

Sweet William, Garden *Dianthus barbatus*

See Carnation for the derivation of the generic name.

Barbatus, meaning "bearded," refers to the appearance of the flowers. The common name, which dates back to the sixteenth century, is of obscure origin.

Sweet William, Wild *Phlox maculata*

See Moss, Pink, for the derivation of the generic and common names.

The species name refers to the purple-spotted stems.

Swiss-cheese Plant *Monstera deliciosa*

The generic name, Latin for "strange" or "monstrous," reflects some of the odd characteristics of

Wild Sweet William

261

this climbing evergreen from tropical America. Among these traits are the long aerial roots, most of which never reach ground; the edible fruits which look like pine cones; and the large leaves, dotted with holes and deep lobes, which account for the common name. The *deliciosa* is applicable to the fruit.

Syngonium *Syngonium podophyllum*
The generic name derives from a technical characteristic of the flower, the cohesion of its ovaries. The species designation means "foot-leaved," an allusion to the shape of the deeply lobed leaves. This climbing or creeping tropical vine is grown in greenhouses and used as a potted plant or climber in homes.

Tamarisk to Twisted Stalk

Tabasco Pepper. See Christmas Pepper

Tamarisk, Kashgar *Tamarix hispida*
This generic name is based on the original habitat of a tamarisk species in the Tamaris River (now Tamber River) region of Spain. *Hispida*, meaning "bristly," refers to finely hairy and feathery leaves. Kashgar is a region near the Caspian Sea where this species was believed to have originated. The tamarisk is a shrub with feathery foliage and fluffy, pink flowers—a really unusual ornamental.

Tangleberry. See Dangleberry

Tansy, Common *Tanacetum vulgare*
Both the generic and common names are rooted in the Greek *athanasia*, "a medicine to prolong life." The Latin *tanasia* was abbreviated to the English tansy. In the Middle Ages and later, the pungent tansy leaves were placed in coffins or rubbed on corpses to preserve them against worms and decay.

Its other uses are many. The tea is used in England to induce menstruation and as a tonic or stimulant. It was used as a wet pack for wounds. The Catawba Indians made a brew from the entire plant which women drank to induce abortions. A powder was used to kill fleas and lice long before the day of chemical insecticides.

Tape Grass. See Eelgrass
Taro. See Dasheen

Tarragon *Artemisia dracunculus*
See Dusty Miller for the derivation of the generic name.
Dracunculus, meaning "like a little dragon," goes back to the Middle Ages or earlier, when many plants, especially those with strong scents or bitter tastes, were associated with belief in dragons. The common name stems from or is related to the Spanish *taragontia* and the Arabic *tarkhon,* both derived from the Latin species name. This anise-scented herb is used as a condiment and as a flavor for vinegar.

Gerard, in his *Herball* is worthy of quotation: "Tarragon is not to be eaten alone in sallades, but joyned with other herbs, as Lettuce, Purslain, and such like, that it may also temper the coldness of them."

Tarweed. See Cuphea, clammy

Tearthumb *Polygonum sagittatum*
See Buckwheat, False, for the derivation of the generic name.
The species name, denoting "arrow-shaped," refers to the leaves. The sharp prickles along the stem can easily tear one's thumb, hence the common name.

Tearthumb

Teasel *Dipsacus sylvestris*
Teasel was once considered a thirst-quencher for the wayfarer who partook of water in the hollows of its leaf bases. The Greek generic name, based on *dipsakos,* meaning "thirst," recalls this use. Teasel's woodland habitat is noted in *sylvestris,* meaning, "growing in the woods."
The common name alludes to an interesting early use of teasel. The dried seed head, covered with firm, hooked bracts, was used to "tease" or card the nap on woollen cloth. These seedheads, dyed, also serve as a winter decoration.

Teasel

Temple Bells *Smithiantha spp.*
This genus honors Matilda Smith (1854–1926),

famed botanical artist at Kew Gardens in London. She also produced drawings for the *Botanical Magazine* from 1878 until 1923. The common name arose from the resemblance of the spike of red bell-like flowers to a set of temple bells. The velvety, red-green foliage as well as the flowers explain the popularity of this genus.

Thimbleweed *Anemone virginiana*

See Pasqueflower for the derivation of the generic name.

The species name indicates that the thimbleweed was first described from a Virginia specimen. The thimble-shaped seed case is dried and dyed as a winter decoration in the home.

Thimbleweed

Thistle, Canada *Cirsium arvense*

See Thistle, Common for generic name.

This thistle, a native of Europe, may have found its way into the United States via Canada and in this way received its common name. *Arvense,* "of the cultivated field," points to this thistle as a serious invader of fields and pastures. The tender stalks, when peeled and boiled, make a very palatable potherb.

Thistle, Common *C. discolor*

This Greek generic name designated a kind of thistle. *Discolor,* meaning "two distinct colors," refers to the typical purple flowers, which are sometimes white.

Thistle is related to the German and Dutch *distel,* meaning "something sharp." It is applied to a variety of related prickly leaved plants and includes the very tall species that is the national emblem of Scotland.

According to an old legend, the Norse invaders of Scotland in the tenth century tried to storm Staines Castle at night. The invaders removed their shoes to wade through the moat, only to find it almost dry and filled with thistles. Their shouts of pain aroused the sleeping garrison, and the invaders withdrew in defeat.

Thoroughwort, Round-leaved *Eupatorium rotundifolium*

See Boneset for the derivation of the generic name.

The specific name is translated in the "round-leaved" portion of the common name. The latter is from an earlier name, thorough wax. The stem appears to grow through the leaves, as in boneset. *Wax* means "to grow," and the *thorough* is an older spelling of the modern "through."

Thyme, Basil *Thymus vulgaris*

Both the common and generic names stem from the Greek *thymon,* "to make a burnt offering or sacrifice," a reference to an ancient role of this herb. *Vulgaris* indicates this herb as common in occurrence. The adjective *basil* means "royal" in Greek and attests to the esteem in which this herb was held.

Garden thyme and allied species are used as seasoning in soups, meats, and other dishes and for an essential oil.

Thyme, Wild *T. serpyllum*

See also Thyme, Basil.

The classical name for wild thyme, *serpyllum,* was bestowed by Virgil and was based on an earlier Greek name. This herb, naturalized in the United States from Southern Europe, is closely allied to basil thyme and has the same aromatic properties and culinary uses. A mixture of thyme and honey was at one time a popular remedy for respiratory ailments. Many horticultural varieties have been developed, including lemon thyme.

Tickseed. See Coreopsis

Tick Trefoil *Desmodium canadense*

The distinctive long pods, divided into joints, each

with one seed, is suggestive of a "long band or chain," the Greek meaning of *Desmodium*. The species name refers to the locale from which the first specimen was described. The common name alludes to three-leaved (trefoil) plants with bristly seeds that readily stick to the clothing of passersby, as does a tick.

Tigerflower *Tigridia pavonia*

The conspicuous spots on the flower explains the generic and common names. The species designation, "as showy as a peacock," gives an indication of the attractiveness of this multicolored tropical flower.

The tigerflower was brought back to Spain in the late sixteenth century by a physician sent to Mexico by Philip II. He reported it as a fever reducer and a herb that would promote fecundity in women. Its popularity today is based solely on its pavonian characteristics.

Tinkersweed; Horse Gentian *Triosteum perfoliatum*

The Greek generic name, meaning "three bones," refers to the three bony seeds usually found in each orange-red berry. The species name indicates that the paired leaves are joined to each other around the stem.

Tinkersweed recalls an otherwise forgotten New England physician of colonial days who promoted the medicinal virtues of this weed, especially its use as a fever reducer and cathartic. Another common name, feverwort, stems from this medicinal use. Horse gentian alludes to its resemblance to a gentian. "Horse" often was applied to a plant resembling another plant, but unrelated or inferior in some way to it.

Known to colonial Germans of Lancaster as wild coffee, the dried toasted berries were considered an excellent substitute for real coffee.

Ti Plant *Cordyline terminalis*

Cordyline, meaning "club-like," is descriptive of

the fleshy roots of this tropical houseplant from the South Pacific. The terminal tufts of elongated, red-maroon leaves explain the species name. The common name is the Tahitian-Samoan word for this decorative plant with green, colored, and variegated leaves. In its native habitat, the roots are used as food and the leaves for thatching and clothing.

Toadflax, Blue *Linaria canadensis*
See Butter-and-eggs for the derivation of the generic name.
Canadensis indicates the locale of the first specimen to be described. Toadflax occurs throughout North America. The prefix "toad" alludes to this plant's resemblance to flax, though it is unrelated.

Toadflax, Yellow. See Butter-and-eggs

Toadshade; Sessile Trillium *Trillium sessile*
Trillium is from Latin *trilix*, "triple," in reference to the whorl of three leaves and the three-petaled flower. *Sessile*, "sitting," is descriptive of the stalkless flowers. One can fancy a toad resting in the shade of this lowly denizen of moist woodlands.
Trillium was used widely by the Indians as a medicinal plant. The mashed roots, steeped in water, were used as a wash for sore nipples, for eye inflammations, and to stop bleeding after childbirth.

Toadshade

Tobacco, Jasmine *Nicotiana alata var. grandiflora*
The generic name recalls Jean Nicot, French consul at Lisbon, Portugal, who sent tobacco seeds to France in 1560. He intended the plant to be used for medicinal purposes rather than for smoking. Both the drug and plant were named in his honor. *Alata*, meaning "winged," refers to the stems. *Grandiflora* is descriptive of the large tubular flowers. These open at dusk, giving off a jasmine-like perfume.

Tobacco, Wild *N. rustica*
See also Tobacco, Jasmine.

Rustica means "pertaining to the countryside," where this species occurs in fields and along roadsides. The closely allied *N. tabacum,* the cultivated tobacco, is believed to be derived from the wild form.

Tomato *Lycopersicum esculentum*
The generic name, signifying "wolf peach," originally applied to an Egyptian plant and was later assigned to the tomato, which once was believed to be poisonous. *Esculentum,* a more modern term, refers to the edibility of the tomato. Thomas Jefferson is quoted in a letter as having stated that "we are indebted to [Dr. John de Sequeyra of Williamsburg, Va.] for the introduction of the admirable vegetable, the tomato. He was of the opinion that the person who should eat a sufficient abundance of these apples would never die."
The common name is traced through the Spanish *tomate* to the Indian Nahuatl word *tomatl.*

Toothwort; Pepper-root *Dentaria diphylla*
The tooth-like projections on the white rootstock account for the generic name and for the old belief that this root was a cure for toothache. The common name, meaning tooth root, alludes to the latter belief. The other common name refers to the peppery flavor of the roots. This edible rootstock can be eaten as a nibble or diced and used in salads. *Diphylla,* "two-leaved" in Latin, refers to the occurrence of the leaves in pairs.

Toothwort

Torch Lily. See Red-hot Poker
Touch-me-not. See Impatience

Trefoil, Birdfoot *Lotus corniculatus*
Lotus, the classical name of several different plants, including the Egyptian lotus, later was applied to the trefoil genus. The species name, meaning "small-horned," probably refers to the small pods. The three leaflets in each leaf and the resemblance of the entire leaf to a bird's foot explain the common name.

Birdfoot Trefoil

269

This member of the clover family bears yellow to red flowers and small, slender pods.

Trillium, Sessile. See Toadshade

Trillium, Tall *Trillium erectum*
   See Toadshade for the derivation of the generic name.
   The red or purple flower is borne on a stalk above the leaves, hence the "erect" appearance and the specific name.

Trout Lily; Adder's-tongue *Erythronium americanum*
   The generic name is based upon the Greek root *erythros,* meaning "red," an allusion to the red European trillium, first in the genus to be described. The species name identifies this one as American. The common name arose because the speckled leaves resemble a speckled trout. The long extruded stamens suggest an adder's tongue.
   The leaves of this streamside herb can be cooked as a vegetable. No harm results where trout lily is abundant.

Trout Lily

Trumpet Creeper; Virginia Creeper *Campsis radicans*
   The name of this genus stems from a Greek word meaning "curve," a reference to the curved stamens. *Radicans* means "rooting on or above the ground."
   The trumpet-shaped orange flowers and the vine's habit of creeping over fences and shrubbery accounts for the common name. The creeper is often cultivated for its attractive flowers and foliage.

Trumpet Creeper

Trumpet Vine, Blood *Phaedranthus buccinatorius*
   The Greek-derived generic name, meaning "splendid-flowered," alludes to the brilliant red color of the trumpet-shaped flowers. The species name is from the Latin *buccina,* "a trumpet." This climbing vine flowers throughout the season.

Tuberose *Polianthes tuberosa*
Polianthes, translated as "white" or "shining," is descriptive of the large attractive white flowers. *Tuberosa* means "tuber-bearing." The tuberose has been grown as a summer garden flower for many centuries because of its beauty and strong fragrance. The precise origin of this species is not known, though related wild species are found in Mexico and Trinidad.

Tulip *Tulipa spp.*
The tulip probably originated in Persia, where its wild progenitor was known as *lale.* The variegated type became known as *tulband* because it resembled a bright-colored turban when inverted.

Turkish travelers to Persia brought back the bulbs of this attractive flower. One wealthy Turk planted a mass of colorful tulips in his palace garden, which became the talk and envy of other merchants in Constantinople. This stimulated a lively business in tulip bulbs. Soon it was commonplace for travelers to Persia to return with bulbs as well as with rugs, perfumes, and brassware.

The Turks, in showing their gardens to European merchants and diplomats, pointed out the resemblance of these flowers to their national headgear, or *tulband.* Visitors took this to be the name of the flower, and it was in time corrupted to *tulipan,* which much later, was abbreviated to tulip. Turkish fondness for tulips grew with the years. It was proclaimed the national flower and an annual feast of tulips was sponsored by the sultan's court.

In 1554, Busbequius, the Austrian Ambassador to Turkey, greatly admired the tulips he saw in Turkish gardens. He paid dearly for some seeds and bulbs, which he sent back to Vienna. Within two decades the tulip had spread to other countries of Western Europe.

An early Dutch tulip enthusiast, Clusius, propagated the bulbs on a large scale and developed new varieties. His work stimulated further interest in this flower in Holland. Some of his best and most promising specimens were stolen from his garden by collectors unwilling to pay the high price demanded. These were propagated and distributed widely. Soon the production of new varieties became a virtual craze.

The famous "Tulipomania" began in 1634 and lasted four years. Prices of scarce and choices bulbs often exceeded the price of precious metals. A single bulb of "Semper Augustus" brought 13,000 florins! After 1638 bulb production and improvement continued, and for over three centuries Holland has been the center of the world's tulip bulb industry.

A milder tulipomania took place in France, where wealthy noblemen and merchants bought and sold bulbs at wild prices. A similar craze occurred in England more than a century after the bubble burst in Holland. Some new varieties developed by British gardeners skyrocketed in price to the equivalent of several hundred dollars a bulb. Tulips reigned supreme in British gardens through the early 1700s.

Today tulips are available in a dazzling array of colors and petal patterns. There are "doubles," "Parrots" with enlarged and frayed petals, and every color except blue. Tulip season now extends for several months, and tulips are available in sizes from one to three feet.

Tumble Mustard

Tumble Mustard *Sisymbrium altissimum*
See Hedge Mustard for the derivation of the generic name.

The species name signifies this as the "tallest" in the genus. At maturity the stem breaks away from the root, causing the plant to be blown about by the wind in a manner similar to the tumbleweed. Since it is related to the mustard, its common name relates to its two significant traits. The tumble mustard makes a good potherb in the spring.

Tumbleweed *Amaranthus albus*
See Amaranth for the derivation of the generic name.

Albus refers to the whitish stems of this weed. On the western plains and in large fields, the dried plants easily break away from their roots in autumn, and their light, rolling mass is driven by winds over prairie and field, until coming to rest along a fence row. This tumbling over the fields is the origin of the common name. Indians and early settlers used the young plants as a cooked green.

Turkey Beard *Xerophyllum asphodeloides*

Xerophyllum, meaning "dry leaf," alludes to the dry, narrow leaves. The species name, meaning "resembling asphodel," is indicative of an actual relationship to that plant. This lily-like pine barrens plant bears a raceme of fragrant, yellow-white flowers, a likely basis for its common name.

Turnip, Garden *Brassica rapa*

See Broccoli for the derivation of the generic name.

Tumbleweed

Rapa is an old generic name, later applied here as a specific name. Turnip is believed to have originated from the Old English *turnep,* based on the French *tour,* "a turn," and the Middle English *nepe.* It frequently escapes from cultivation and can become a serious pest.

Turtlehead *Chelone glabra*

Both the generic and common names are based upon the fancied resemblance of this flower to a turtle's head with the mouth open. *Glabra,* or "smooth," is descriptive of the smooth surface of leaves and stems. Several varieties under cultivation bear pink, white, and purple flowers.

Twayblade *Liparis lillifolia*

The Greek word *liparis,* meaning "oily or smooth," refers to the smooth leaves of this orchid. *Lillifolia,* Turtlehead or "lily-like leaves," further describes the foliage. *Twayblade* is an old English word for two leaves or blades. There is only one pair of leaves per plant.

Twinflower *Linnaea borealis*

Linnaeus, in naming this genus for himself, described it as "a plant of Lappland, lowly, insignificant, flowering for but a brief space—from Linnaeus, who resembles it." Linnaeus' *Genera Plantarum* is the starting point of modern systematic botany. An interesting addendum is that Linnaeus posed for his portrait with a sprig of twinflower in his hand. *Borealis,* "of the north," refers to the habitat of twinflower, the northern half of the United States.

There are always two flowers on each stalk, hence the common name.

Twinleaf *Jeffersonia diphylla*
This genus was named in honor of Thomas Jefferson, in recognition of his status as a horticulturist, patron of botany, and an advocate of improved farming methods. *Diphylla,* meaning "two-leaved," alludes to each leaf being almost divided in two to give the outward appearance of two leaves.

Twinleaf

Twisted Stalk *Streptopus roseus*
The Greek generic name, translated as "twisted foot," is descriptive of the twisted flower stalk, which zig-zags at each leaf joint. *Roseus* refers to the rose-colored flowers in each leaf axil. The red berries, eaten by rural folks, should be consumed in moderation as they are somewhat cathartic.

Twisted Stalk

274

Umbrella Plant to Unicorn Plant

Umbrella Plant *Cyperus alternifolius*
Cyperus is the ancient Greek name of this ornamental sedge. *Alternifolia* alludes to the alternate arrangement of the grass-like leaves. Eight to twelve linear leaves are attached to the stem at one point, forming an umbrella, hence the common name.

Umbrellawort *Allionia nyctaginea*
Carlo Allioni (1725–1804) was professor of botany at Turin University in Italy. He wrote extensively on the plants of the Italian Piedmont, publishing a three-volume work on the subject and describing over 2,800 species. He was Director of the Botanical Gardens of Turin and won recognition by his election as a member of the Royal Society in London. The species name signifies "night-blooming." The common name is descriptive of the umbrella shape of the flower head.

Unicorn Plant *Proboscidea louisianica*
"Snout-like" is the translation of the generic name, a reference to the appearance of the six-inch long pod which splits in two, each part bearing a hooked horn or curving beak. Its odd pod also gave rise to the fanciful common name. This southern plant was first described from a Louisiana specimen, hence the species name. This bushy, sticky plant bears yellow flowers with purple spots and is sometimes cultivated.

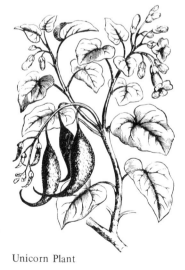

Unicorn Plant

Valerian to Virginia Creeper

Valerian

Valerian; Garden Heliotrope *Valeriana officinalis*
The Latin *valere,* meaning "to be strong and healthy," probably was chosen as a generic name for this group in recognition of its powerful medicinal properties, especially for hysteria and nervousness. The species name alludes to its place in the apothecary shop.
The alternate common name arises from a resemblance of this plant to the heliotrope.

Valerian, Greek; Jacob's Ladder *Polemonium reptans*
See Jacob's Ladder for the derivation of the generic name.
Reptans means "creeping," apt for a plant which seldom exceeds a foot in height. The common name helps to distinguish this from the preceding unrelated species. The word "Greek" crept into the name because the generic name is a well-known Greek plant name.

Vanda *Vanda spp.*
This generic name stems from the Hindu *vanda,* a mistletoe-like plant, and from the same word in Sanskrit, meaning a parasitic plant. This epiphytic orchid, native of Asia, bears yellow-brown and lavender-blue flowers on long racemes. Vandas are climbing vines which develop into true epiphytes high in the trees; they are not parasitic.

Vegetable Marrow *Cucurbita pepo*
See Gourd for the derivation of the generic name.
Marrow is a form of squash and pumpkin, both of which are *C. pepo. Marrow* signifies the choicest

of food, the seat of animal vigor or health. In horti-
culture it is the choice pulp of a vegetable or fruit.
This egg-shaped gourd is eaten as a vegetable in
England; it is less popular in the United States.

Velvet Flower; Painted Tongue *Salpiglossus sinuata*
Salpiglossus, Greek for "scalloped-tube tongue,"
refers to the tongue-like style in the flower tube.
Sinuata, or "wavy-margined," refers to the leaves.
The velvety appearance of this trumpet-shaped
flower explains its common name. Painted tongue
refers to the style mentioned above.

Velvetleaf; Butterprint; Piemarker *Abutilon theophrasti*
Abutilon is the Arabic name given by Avicenna to
this mallow-like plant. The species name honors
Theophrastus, Greek philosopher and naturalist.
   This plant bears velvety, heart-shaped leaves,
which accounts for its common name. This leaf was
often used by the farmer's wife as a butterprint or to
mark pies.

Velvetleaf

Velvet Plant *Gynura aurantiaca*
"Tailed stigma" is the translation of the Greek
generic name, descriptive of this flower's odd stigma.
The species name alludes to its yellow or orange
flowers. The succulent stems and leaves, densely covered with violet hairs, account for the origin of the
common name.

Venus's-flytrap *Dionaea muscipula*
Dionaea, the Greek name of the goddess Venus,
was bestowed upon this insectivorous member of the
sundew family, apparently for capricious reasons.
The species name, meaning "fly-catching," is an accurate description of the feeding habits of this unusual plant. The flytrap occurs only in the Carolina
swamps, though it is offered in the catalogs of a
number of nurserymen. The leaves are arranged in a
basal rosette.
   Each leaf has two hinged blades with long hairs

277

on the outer edges and three tiny trigger hairs on the inner surface. When these are touched by alighting insects in search of nectar, the two blades fold together tightly, trapping the insect. The trapped fly or gnat dies, and it is digested by the plant.

Venus' Looking-glass *Specularia perfoliata*

The generic name is based upon an ancient Latin name *speculum veneri,* "mirror of Venus." The species name alludes to the leaf which is pierced by the stem or surrounds it.

Verbena, Garden *Verbena venosa*

This ancient Greek name was given by Pliny. This plant was considered an herb sacred to Dioscorides and was used to cleanse the festive tables. *Venosa,* meaning "veined," refers to the prominent leaf venation of this summer bedding plant.

Verbena, Lemon *Aloysia triphylla*

The lemon verbena was named for Maria Louisa Theresa, wife of Charles IV of Spain. She was a strong influence in his court during the two decades of his reign. The botanist who honored the queen chose to Latinize her middle name, Louise. A tea made of the dried leaves of this herb was popular in Spain at the time, possibly a factor in the selection of the generic name.

Venus' Looking-glass

The species name relates to the leaves, which are borne in threes. The lemon verbena, a native of South America, was introduced into North America in the early 1700s. It is still popular as a houseplant because of the strong, pleasing aroma.

Vervain, Blue *Verbena hastata*

See Verbena, Garden for derivation of the generic names.

Hastata, signifying "spear-shaped," is descriptive of the leaf shape of this wild species. Vervain is a corruption of the generic name.

The Druids of ancient Britain used vervain in medicine and held it in reverence because of the

Blue Vervain

278

resemblance of its leaves to those of certain oaks. The Crusaders brought back tales of this marvelous plant found growing on Mt. Calvary and of its miraculous cures of the sick. A European species is known as *V. officinalis* because it was commonly sold in the apothecary shop.

Today verbena has no medicinal standing; verbena oil, also called lemon-grass oil, has a limited use in perfumery.

Vetch, Hairy *Vicia villosa*

See Bean, Yard-long, for the derivation of the generic name.

The pods, covered with "soft hairs," account for the species name. Vetch is simply a corruption of the generic name. Vetches are widely used as fodder crops, for soil improvement, and as a ground cover. The broad bean, *V. faba,* belongs to this genus.

Vetchling; Beach Pea *Lathyrus palustris*

See Beach Pea for the derivation of the generic name.

Palustris, or "marsh-loving," alludes to the usual habitat of this species. The seeds of vetchling were once reputed to possess aphrodisiacal properties. They can cause temporary spastic paralysis of the legs if eaten in any quantity.

Vetchling

Violet *Viola spp.*

According to Greek legend, Zeus created the violet as a special sweet-smelling food for Io, daughter of the river god Inarchus. *Io* is Greek for "violet," and the English name derives from the Greek, through the Latin *viola* and the French *violette.*

This popular woodland denizen is the state flower of Illinois, New Jersey, Rhode Island, and Wisconsin. There are about 300 species world-wide; 80 grow in the United States. Violets occur in many colors: purple, yellow, cream, white, and green, as well as many shades of blue. There are also several two-colored species. Well-known cultivated forms are viola and pansy.

Violets produce two kinds of flowers to insure survival of the species. The showy spring flowers are cross-pollinated by insects in search of the nectar. In summer petal-less, closed flowers are borne on short stems close to the ground. These are self-pollinated and produce good seed, an assurance of survival in the event the spring flowers were not pollinated because of rain or cold weather.

The problem of identification has perplexed botanists for years. With the assistance of insects, almost all species hybridize readily. If a plant found in the woods has traits intermediate between two recognized species, it is a probable hybrid. It might also be a hybrid of a hybrid, with possibly four species involved in its ancestry. The avid violet hunter is likely to find albinos. One well-known type is the Confederate violet, which is white with a blue center.

Violets long have had culinary and medicinal uses. Pliny tells us that in Greece, a violet concoction induced sleep for the insomniac and strengthened the heart of the cardiac. In pre-Roman Britain women made an infusion of violets in goat's milk, which they applied to the face as an aid to the complexion. Through the centuries violets have been used in the treatment of many ailments.

In modern times both leaves and flowers have culinary uses. Violet leaves, which contain Vitamins C and A, are popular as a potherb, prepared like spinach or cut up finely in a salad. Go easy, since these leaves are mildly laxative to some. Known locally as "wild okra," violet leaves are used to thicken soups.

Violet flower heads can be candied or made into jam, jelly, or syrup. To pick violets, run your open fingers through a clump and break off several flowers in one sweep. To make violet jam, blend to a paste of one part violets, one part water, and three parts sugar, plus a tablespoon of lemon juice. Dissolve a packet of pectin in water, bring to a boil, and pour into violet paste, blending for a few minutes. Pour into containers and refrigerate. Use within a month.

Candied violets are a gourmet's delight. Dissolve an ounce of gum arabic in a half cup of water. Let

Violet

it cool, and brush the flowers with the dissolved gum. Let dry, dip in a syrup made of corn syrup and sugar, and sprinkle with confectioner's sugar. Place this delicacy on waxed paper sheets to dry. The finished product can be used as a topping for cakes and other desserts.

Violet jelly is a delightful breakfast spread that is easy to prepare. Pack a glass container with blossoms, pour on boiling water, and let stand for 24 hours. Bring to a boil two cups of the strained infusion and add a packet of pectin plus two tablespoons of lemon juice. Now add four cups of sugar. Boil vigorously for a minute, pour into containers, and seal.

Violet, Bird's-foot; Wild Pansy *V. pedata*
See Violet for the derivation of the generic name.
The leaves are cleft into segments resembling a bird's foot, hence the common and specific names. The upper petals are purple, the lower lilac, which is the basis for the alternate name of wild pansy.

Bird's-foot Violet

Violet, Canada *V. canadensis*
See Violet for the derivation of the generic name.
This species was first described from a Canadian specimen, hence the species and common names. It is common throughout the United States, except the far South.

Violet, Cream *V. striata*
See Violet for the derivation of the generic name.
Striata, meaning "striped," refers to the fine purple lines on the cream-colored petals.

Violet, Green *Hybanthus concolor*
This is closely allied to the *Violas,* but differs in the presence of a dorsal swelling on the flower, the meaning of the Greek generic name. *Concolor,* or "one-colored," refers to the one color of flower and leaves. This violet is sometimes hard to find because of its green flowers.

Smooth Yellow Violet

Violet, Smooth Yellow *V. pensylvanica*
See Violet for the derivation of the generic name. This common yellow violet was first described from Pennsylvania, though its range is from Quebec to Georgia.

Violet, Sweet White *V. blanda*
See Violet for the derivation of the generic name. This specific name, denoting "alluring" or "tempting," alludes to the large size and fragrance of this violet.

Viper's Bugloss *Echium vulgare*
Echium, the original Greek name for this plant, was based upon the root *echis,* signifying "viper." *Vulgare,* "common," signifies the common occurrence of this weed, also known as blueweed.

The common name stems from the Latin *buglossa* and the Greek *bouglossos,* both meaning "ox tongue" and both descriptive of the fanciful appearance of the flower, an open mouth with a protruding tongue. Because of their supposed resemblance to the head of a viper, these seeds were thought to be an effective remedy for a snakebite. The prescription called for eating the root while imbibing wine.

Virginia Creeper

Virginia Creeper *Parthenocissus quinquefolia*
The old French name *vigne-vierge* was adopted as the generic name of this vine and translated to Greek as *parthenocissus,* the "virgin's ivy." The common name is based on its creeping or climbing habit and on its common occurrence in Virginia.

Quinquefolia, "five leaves," refers to the five-parted leaves of this species. The leaves turn a brilliant scarlet in the fall, and the blue berries are attractive to birds. There are several cultivated varieties, including the silver-vein creeper and Japanese ivy.

Virginia Snakeroot. See Birthwort

Wallflower to Wormwood

Wallflower *Cheiranthus cheiri*

Two Greek words meaning "hand" and "flower" comprise the generic name, which is based on an old-fashioned custom of carrying these scented flowers by hand as a gift bouquet. The species name repeats the "hand" portion of the generic name, for reasons not clear today.

These sweet-scented flowers are commonly found in Europe growing on old stone walls, rocks, and cliffs, hence the common name.

Water Arum; Wild Calla *Calla palustris*

Calla is derived from the Greek word "beautiful." This seems to be a misapplication, since this flower cannot be classed as "beautiful," nor is it related to the attractive calla lily. *Palustris* refers to the swamp habitat of this arum.

The common name indicates kinship with the ancient arums, a name which may be derived from Aaron of the Bible, after whom several plants were named. In Lapland and elsewhere a bread is made of the meal derived from the starchy rootstock. It takes considerable digging, grinding, and boiling to produce a loaf.

Water Arum

Water Chestnut *Trapa natans*

This Greek generic name is derived from *calcitrapa*, or "caltrop," because of the spreading, sharp points of the hard seed-bur. The calcitrapa was a Roman defensive war weapon made of four sharp iron points. It was placed in the ground in the path of cavalry to pierce hooves and disable horses. *Natans*, meaning "swimming," refers to the aquatic habitat of this herb.

283

The water chestnut is so-named because of its habitat and the spiny, nut-like fruits, suggestive of a chestnut. The seeds are edible when roasted and can be candied like the true chestnut. This European plant has spread widely in the southeast quadrant of this country, where it tends to clog streams.

Watercress *Nasturtium officinale*
The pungent flavor of this cress is the basis for the unusual generic name, which is Latin for "twisted nose." It is difficult to imagine this salad and mild condiment causing a nasal spasm. The species name indicates its place in the market or apothecary shop.

Cress is an old plant name, stemming from the Middle English *cresse* and the Anglo-Saxon *cresse* and *cerse*. Watercress is found in clear running or spring waters.

After several centuries of identification with watercress, the popular name nasturtium was transferred to a garden flower, the South American *Tropaeolum*. The switch occurred because of the similar taste of the leaves.

Watercress

Water Hemlock *Cicuta maculata*
Cicuta is the classical name for poison hemlock, the poison taken by Socrates. The species name, meaning "spotted," refers to an identifying characteristic, the stem streaked and spotted with purple. Hemlock is derived from the Anglo-Saxon *hemlic,* the general name for poisonous herbs.

Water Hemp *Acnida cannabina*
Acnida, meaning "stingless," was applied to this genus to distinguish it from a stinging plant which it resembled. *Cannabina* means "resembling hemp," which this does. The water hemp inhabits coastal salt marshes.

Water Hemlock

Water Horehound; Bugleweed *Lycopus americanus*
See Horehound for the derivation of the common name.

The leaves of several species are fancied to resemble a wolf's foot, the meaning of the generic name. *Americanus* distinguishes this as a North American species. Bugleweed and bugle probably have similar origins. See the former for a discussion of its origin.

The white tubers, boiled in salt water, are an agreeable vegetable. These also can be pickled or cut into small pieces and used in a soup or stew.

Water Hyacinth *Eichhornia crassipes*
J. A. Eichhorn (1779–1856), a Prussian patron of horticulture, is recalled through this genus. *Crassipes,* "thick-footed" or "thick-stemmed," is descriptive of this plant's stalks.

The flowers of this aquatic bear a slight resemblance to the hyacinth. This tropical American native, introduced in the twentieth century, has now spread through a large area of the Southeast, often clogging sluggish streams and ponds. Leaves and stems can be eaten, cooked, or steamed as a palatable green vegetable.

Water Hyacinth

Water Lily, Fragrant *Nymphaea odorata*
This genus honors the nymphs of Greek mythology who dwelt in seas and streams, the habitat of the water lilies. *Odorata* refers to the fragrant flowers. The common name arose from the resemblance of this large white flower to that of a lily.

The unopened flower buds, cooked a few minutes in boiling water and properly seasoned, make a unique, palatable dish. The young, unfurled leaves, chopped up and boiled for several minutes, make a desirable addition to soups and stews.

Watermat. See Golden Saxifrage

Watermelon *Citrullus vulgaris*
Citrullus is the classical Latin name for watermelon. *Vulgaris* refers to its common occurrence in its South African desert habitat. The common name

285

arose from the copious, sweet watery juice of this melon. David Livingstone, the African explorer, described the watermelon as abundant in the Kalahari Desert. It was introduced into India at an early date, since it has a Sanskrit name, and it reached China in the tenth century. It was not known to the Greeks and Romans until after the beginning of the Christian era.

Broadleaf Water Milfoil

Water Milfoil, Broadleaf *Myriophyllum heterophyllum*
The generic name, Greek for "myriad of leaves," alludes to the extremely divided, feathery foliage. The species name, "diversely leaved," refers to the two types of leaves; that is, the submerged foliage described above, and the emersed leaves, which are large and lance-shaped to elliptic. The common name also deals with the leaf characteristics; the *milfoil* or "thousand leaves," allude to the submerged foliage, the "broad leaf" to the leaves on or above the surface.

Water Primrose; Primrose-willow *Jussiaea repens*
Bernard de Jussieu (1699–1777) was the French botanist who laid the foundation for a system of plant classification. He was professor at the Royal Gardens in Paris (1722) and Superintendent of the Royal Gardens of Louis XV (1758). Linnaeus named this genus in Jussieu's honor in 1737, in gratitude for his pioneering efforts toward a system of grouping plants.
Repens, meaning "creeping," identifies the growth habit of this denizen of swamps and shallow waters.
This herb is neither a willow nor a primrose. The name apparently arose from supposed resemblances of leaves or flowers.

Water Purslane *Ludwigia palustris*
See Seedbox for the derivation of the generic name.
Palustris, "swamp-dwelling," indicates the semi-aquatic habitat of this species. This prostrate or floating plant has a superficial resemblance to the

true purslane, hence the common name.

Water Shield *Brasenia shreberi*
This Latin generic name is of unknown origin.
The species name honors Johann C. D. Shreber
(1793–1810), a noted German botanist. The water
shield is an aquatic plant with floating, shield-shaped
leaves, somewhat like the water lily. The Indians
relished the starchy tuberous roots.

Water Shield

Water Speedwell. See Brook Pimpernel

Water Willow *Decodon verticillatus*
The ten points or "teeth" of the calyx account for
the generic name, Greek for "ten teeth." The species name, "whorled," denotes the typical three
leaves in a whorl around the stem. This aquatic herb
with willow-like leaves indicates the origin of the
common name. Water willow is found in swamps
and along edges of pools and slow streams.

Wax Plant *Hoya carnosa*
This genus was named in honor of Thomas Hoy,
gardener to the Duke of Northumberland. *Carnosa,*
Latin for "fleshy," refers to the unusually thick
leaves. The wax plant bears fragrant, waxy-white
flowers with pink centers. It is a twining plant, supported by aerial roots.

Waxweed, Blue. See Cuphea, clammy

Weigela, Old-fashioned *Weigela florida*
This genus honors C. E. Weigel (1748–1831), a
Swiss physician and botanist. *Florida* refers to the
attractive flowers for· which this shrub is known.
There are varieties with red, white, pink, and yellow
flowers. Some white-flowered varieties are at first
pure white but gradually change to pink.

Old-fashioned Weigela

White Campion. See Lychnis, Evening

Whitlow Grass *Draba verna*
Draba, Greek for "acid," refers to the acidic taste of the leaves. *Verna,* "spring," informs us of the season of flowering. See Whitlow-wort for an explanation of the word "whitlow."

Whitlow Grass

Whitlow-wort, Silver *Paronychia argyrocoma*
The generic name is Greek for "whitlow," a term little used today. It means a sore at the quick, beneath a fingernail or under a horse's hoof. It also can signify a swelling of the finger or thumb. A whitlow-wort is a plant which has supposed curative properties when applied to such sores.
The species name is translated "silver-haired," a reference to the silvery bracts.

Wild Basil *Satureia vulgaris*
See Savory, Summer, for the derivation of the generic name.
Vulgaris, "common," alludes to the widespread occurrence of this aromatic herb. *Basil* is Greek for "royal," a reference to use of this herb in the royal bath.
Wild basil goes well with tomato-based dishes, and the chopped leaves are used in egg and fish dishes and spaghetti sauce. The flowers are very attractive to bees. Basil tea is a home remedy for sore throat.

Silver Whitlow-wort

Wild Ginger *Asarum canadense*
Asarum is a classical plant name, latterly applied to wild ginger. The species name indicates that it was first described from a Canadian specimen. In the United States it occurs as far south as the Carolinas. Ginger traces to the Old English *zingiber,* which is from the Latin *gingiber.*
The ginger aroma of the rootstock gave rise to the common name. It grows close to the surface and can be easily collected. The thick root can be cut up,

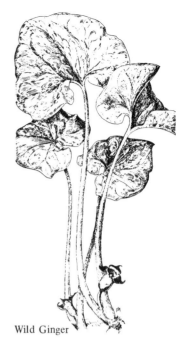

boiled to tenderize, and cooked in a heavy sugar syrup to make candied ginger. It also is used to flavor cookies and cakes. It must be dried slowly, chopped fine, and mixed into the dough.

The American Indians used wild ginger as a contraceptive. The root was boiled slowly in a small quantity of water, and the bitter liquid taken as a hot tea. There is no record as to its effectiveness.

Wild Quinine; American Feverfew *Parthenium integrifolium*

The generic name, originally applied to a plant with white ray flowers, was later bestowed on this group. The Greek name means "maiden's flower." This plant once was used widely as a fever reducer and tonic, hence the two common names. Feverfew is derived from the Latin words *febris* and *fugare,* meaning "fever" and "put to flight." *Integrifolium* refers to the entire, unlobed leaves.

Wild Ginger

Willow Herb, Purple-leaved *Epilobium coloratum*

See Fireweed for the derivation of the generic name.

The species name means "colored," a reference to the bright red flowerstalk. The common name is descriptive of the purplish, narrow, willow-like leaves. The vigorous, young shoots make an excellent substitute for asparagus. The young leafy stems are good as a cooked vegetable. In England the dried leaves serve as a tea substitute.

Purple-leaved Willow Herb

Windflower *Anemone quinquefolia*

See *Pasqueflower* for the derivation of the generic name.

The specific name, meaning "five-leaved," is a misnomer; the single leaf is deeply divided, giving the appearance of five leaves.

The ancients believed that this flower did not open until stirred up by the winds. Another legend is that the windflower's original home was on the slopes of Mount Olympus, the mountain of the gods, where prevailing winds blow. The windflower has five petal-

Windflower

289

like white sepals, that look like petals to the non-botanist.

Wineberry *Rubus phoenicolasius*
See Blackberry for the derivation of the generic name.
The species name indicates that this plant originated in ancient Phoenicia (modern Syria and Lebanon). The common name reflects the popular use of these berries in wine-making. They also can be eaten fresh, cooked, and preserved.

Wineberry

Wingstem *Actinomeris alternifolia*
The irregularity of ray flowers in some species—a highly technical matter—accounts for the Greek name, meaning "ray" and "part." The alternate arrangement of leaves on the stem is reflected in the specific name. The common name alludes to the longitudinal "wings" along the stem, as in many sunflowers.

Winter Aconite *Eranthis hyemalis*
The Greek generic name proclaims this as one of the earliest flowering herbs. The species name, meaning "of the winter," indicates that the basal leaves live through the winter. This tuberous-rooted garden perennial bears bright, buttercup-like flowers.

Winterberry; Black Alder *Ilex verticillata*
Ilex derives from the resemblance of the leaves to those of Virgil's holm oak. The whorled leaf arrangement is the meaning of the species name. This holly species produces bright red berries that persist through the winter, hence the common name.
Alder derives from Middle English *aller,* Anglo-Saxon *alor,* and Latin *alnus.* The Black Alder was once thought to be allied to the Alder.
The dried leaves, collected in summer, are used in preparing a tea. The berries, however, are a strong cathartic and can cause vomiting.

Winter Purslane *Montia perfoliata*

Guiseppe Monti, in whose honor this genus was named, was a professor of botany in Bologna University in the first half of the eighteenth century. The species name alludes to a remarkable feature, a cup one or more inches across, from which arises the racemes of the small white flowers. This fleshy-leafed herb of the portulaca family can be used as a salad herb.

Wishbone Flower *Torenia fournieri*

The Rev. Olaf Toren (1718–1753), recalled by the genus, was chaplain to the Swedish East India Company trading stations in India and China. A plant collector during spare time, he discovered *Torenia asiatica,* which he introduced to Europe. The species name honors Eugene P. N. Fournier (1834–1884), a French botanist.

The color pattern of the flower resembles a wishbone, hence the common name. This window box or border plant produces flowers all through the summer, until the frost. It takes the place of the pansy in the deep South.

Wisteria, Chinese *Wisteria sinensis*

This genus was dedicated to Caspar Wistar (1761–1818), professor of anatomy at the University of Pennsylvania College of Medicine and notable scientist. He succeeded Thomas Jefferson as president of the American Philosophical Society. The common name is derived from the scientific name.

This twining vine of the pea family often climbs up trees, on trellises, or around porches. It bears aromatic, violet-blue flower clusters. The Japanese species, *W. floribunda,* bears smaller but more fragrant flowers.

Chinese Wisteria

Witch Hazel *Hamamelis virginiana*

This generic name, based on the Greek words meaning "fruit" and "at the same time," is easily explained. The yellow flowers and the previous season's seed capsules are on the bush at the same time, often in midwinter during spells of mild weather. *Virginiana* refers to the state from which the species was described.

The common name has an involved but interesting origin. *Witch* is from the Anglo-Saxon *wicen,* "to bend," and refers to the pliant quality of the stem. Hazel wood was used in making divining rods. Thus, the original European shrub was long associated in popular lore with divining and with uncanny phenomena. In America the name was transferred to a different shrub, the stems of which also were used in making divining rods.

The American Indians used the dried leaves in preparing a tea. A strong decoction of the leaves and bark was used as a liniment for muscular aches and as a wash for skin and eye inflammations and for bruises.

Woadwaxen *Genista spp.*

See Dyer's Greenwood for the derivation of the generic name.

Several species of woadwaxen are native of Europe and grow in poor soil and sunny locations. They are small deciduous shrubs, often with minute leaves. The green branches take the place of leaves in food manufacturing.

The common name stems from the Middle English *wod,* the Anglo-Saxon *wad,* the German *waid,* and possible to the Latin *vitrum,* a plant producing a blue dye. The second syllable is from the word *wax,* "to grow." A dyestuff is prepared from the leaves of woads.

Wood Betony *Pedicularis canadensis*

See Lousewort for the derivation of the generic name.

The species name denotes Canada as the locale from which the species was first described. *Betony* is derived from the Latin *betonica,* a plant name stemming from *Vettones,* a Spanish tribe with which this herb was associated in Roman times. Also known as lousewort, this species occurs in dry woods.

Wood Betony

Wood Mint *Blephilia ciliata*

This Greek generic name means "eyelash," from the resemblance of the cilia on the flowers to eyelashes. The species name also refers to cilia that fringe the flowers. The common name identifies this as a woods-dwelling mint.

Woodruff, Sweet *Asperula odorata*

Asperula is Latin for "slightly harsh" or "rough," an allusion to the bristly leaves and stem. The aromatic quality of this herb is noted in the species name. *Woodruff,* an old plant name, stems from the Anglo-Saxon *wudurofe.* This herb, introduced from Europe, is locally established and has some popularity in gardens. It is used as a flavor in wine and liqueur and to some extent in perfumery.

Wood Sage. See Germander

Wood Sorrel, Violet *Oxalis violacea*

See Oxalis for the derivation of the generic name. The violet-colored flowers explain the specific name. The origin of the common name is found under *Oxalis.* This immigrant from Europe is widely naturalized in the United States.

Wood Sorrel, Yellow *O. stricta*

See also Wood Sorrel, Violet.

Stricta, meaning "upright," refers to the growth habit of this species, a plant sometimes identified as the shamrock of Ireland. The other shamrock is the white clover.

This plant has clover-like leaves, but the flowers have five petals. The tender, acid leaves are refreshing to the taste and make an interesting addition to a salad. This wood sorrel, native of Europe, is found throughout temperate North America.

Wormseed *Chenopodium ambrosioides*

The species name means "resembling ambrosia," Yellow Wood Sorrel

293

Wormseed

the ragweed group. There is a superficial resemblance in the plant shape and the deeply incised leaves of both groups.

The anthelmintic (worm-removing) properties of the seeds of this herb have been known since the time of the Maya Indians of Mexico. Wormseed was in the U.S. Pharmacopoeia from 1820 until 1947, when it was replaced by a synthetic product.

Wormwood *Artemisia absinthium*

See Dusty Miller for the derivation of the generic name.

The common name refers to absinthe, a green alcoholic beverage containing oils of wormwood, anise, and other aromatic herbs, widely used in France until prohibited in 1915, and less popular though widely used in other countries. It is more intoxicating than some other liquors, and if taken in excess, it causes a nervous derangement known as absinthism.

Wormwood derives from or is related to the Anglo-Saxon *wermod,* the German *wermuth,* and the French *vermouth.* Vermouth today is a dry white wine, flavored with aromatic herbs. Wormwood also was used "straight," as a tonic and anthelmintic.

Woundwort. See Hedge Nettle

Yam to Yucca

Yam. See Sweet Potato

Yam, Chinese; Cinnamon Vine *Dioscorea batatas*
This genus honors the first-century Greek physician Dioscorides, author of *Materia Medica,* which described the medicinal herbs known in his time. *Batatas,* meaning "potatoes," refers to its edible tubers. *Yam* is an abbreviation and corruption of the African vernacular words *inhame* and *nyami,* basically meaning "to eat," and applied specifically to the starchy, tuberous roots of this genus.
This twining vine, bearing cinnamon-scented flowers, can be grown as far north as New England. In the South it often escapes from cultivation. Its tubers sometimes grow two to three feet long.

Yarrow *Achillea millefolium*
The Greek generic name honors Achilles, hero of the Trojan Wars. He learned of this plant's properties in promoting the healing of wounds from Cheiron the centaur. *Millefolium,* literally "a thousand leaves," is descriptive of the feathery appearance of the much-divided leaves.
Yarrow stems from the Middle English *yarowe,* the Anglo-Saxon *gearwe,* and the German *garbe.* It is a strongly scented herb, widely naturalized in North America. During the Civil War it was known as soldier's woundwort. The leaves were pulverized, steeped in boiling water, and applied to wounds. Earlier the American Indians used it in the same manner, but extended its use to the treatment of burns, earaches, and bruises.

Yarrow

295

Yellow Cress *Rorippa islandica*
The Latin *roro,* "to be moist," and *ripa,* "river-bank," clearly define shallow waters and moist places as the habitat of this genus.
The species name identifies the supposed original Icelandic habitat. The common name alludes to the cress-like appearance of the plant. The young shoots of yellow cress are edible in a salad or as a potherb.

Yellow Cress

Yellow-eye Grass *Xyris caroliniana*
Xyris, Greek for "sharp," the name. used by Dioscorides for an iris, was later applied to this genus. The specific name denotes the habitat of the first specimen to be described. The common name alludes to the flower color and the grass-like foliage. Xyris is found in bogs, pine barrens, and sandy swamps. The leaves and roots were once used in the treatment of skin diseases.

Yellow Flag *Iris pseudoacorus*
See Blue Flag for the derivation of the generic name.

Yellow-eye Grass

The specific name, "false acorus," indicates a superficial resemblance, but no relationship, between this iris and *Acorus calamus.* The common name was suggested by the predominantly yellow flower, which looks like a waving flag on a long, erect stalk.

Yucca *Yucca filamentosa*
Both the generic and common names stem from the Spanish *yuca,* which in turn is based on the similar Taino Indian word for this plant. The loose threads or fibers on the leaf margin account for the species name, meaning "filamentous." This leaf fiber was used widely by the Indians of the Southwest. Also known as Spanish bayonet, the yucca is now a common escape from cultivation. The fresh flowers, properly dressed, have been used in salads. Yucca is the state flower of New Mexico.

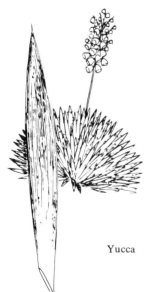

Yucca

Zebra Plant to Zinnia

Zebra Plant *Calathea zebrina*
See Calathea for the derivation of the generic name.
This perennial foliage plant bears very striking striped leaves, suggestive of the zebra's coat; hence the species and common names. The alternate bars are a rich velvety green and a pale yellow-green. The undersides of the leaves are purple-red.

Zenobia, Dusty *Zenobia pulverulenta*
Zenobia Septimia was queen of Palmyra, A.D. 267–272. This city, near present-day Damascus, was believed to have been built by King Solomon on the trade route from Persia to Egypt. Zenobia succeeded King Odenathus to the throne upon his death in A.D. 266. She promptly declared her city-state independent of Rome. On learning this rash act, Emperor Aurelian's troops stormed and captured the city and took the queen prisoner. After an abortive revolt in A.D. 273, the Roman general ordered the city destroyed and reduced to rubble and dust. This "dust-covered" genus Zenobia recalls the fate of the bold queen's city-state, long covered by dust and sand.
The species name, meaning "powder- or dust-covered" also is descriptive of the gray-green foliage of *Zenobia*. This relative of the blueberry has white bell-shaped flowers.

Zephyr-lily *Zephyranthes atamasco*
The fanciful generic name means "flower of the west wind," an allusion to its habitat in the Western Hemisphere. *Atamasco* is the Indian word meaning

Zephyr-lily

297

"stained with red," possibly a reference to the pink flowers.

This relative of the amaryllis bears white or pink flowers and has grass-like leaves. Moist meadows are its preferred habitat.

Zinnia *Zinnia elegans*

Zinnia

Johann G. Zinn (1727–1759) was simultaneously a professor of medicine and director of the Botanical Gardens of the University of Gottingen (1753–1759). He wrote a descriptive catalog of the plants of the Gottingen region in 1757 and *Anatomy of the Eye,* the first book on this subject. Linnaeus named this genus after Zinn in recognition of his work in botany.

"Elegans" signifies the elegance of one of many of the hybrid zinnias developed from the wild species of Mexico and the southwestern United States. Breeders have produced zinnias in every color except blue and true pink, and in size they range from four-inch dwarfs to three-foot giants.

Glossary

Bibliography

Index

Glossary

ACUMINATE: a long-tapering pointed end, as of a leaf.

ALTERNATE: not opposite; e.g., as leaves arising singly at different points on a stalk.

ANNUAL: a plant that lives for only one season.

ANTHER: the pollen-bearing part of the stamen.

AXIL: the angle between the main stalk and a leaf twig or branch arising from the stalk.

BEARD: a clump of hairs in a tuft over a small area.

BERRY: a fruit with fleshy pulp surrounding the seeds.

BIENNIAL: a plant that matures, produces seed, and dies in its second season.

BRACT: a modified, often colorful leaf beneath or surrounding a usually inconspicuous flower or flower cluster.

BRISTLE: a stiff hair on a leaf, stem, or other plant part.

CALYX: the outer parts of a flower, usually sepals, as distinct from the inner parts or petals.

CAMPANULATE: bell-shaped (as of flowers).

CAPSULE: the dry container of seeds.

CLEISTOGAMOUS: producing seed from self-fertilized flowers; the flower does not open.

COMPOUND LEAF: a leaf divided into several leaflets.

CORDATE: heart-shaped, as a leaf.

CORM: an enlarged solid fleshy bulb-like base of a stem.

COROLLA: the inner parts of a flower, usually petals, as distinct from the outer parts or sepals.

CORYMB: a flat-topped open flower cluster, the outer flowers blooming earliest.

DECIDUOUS: with leaves falling in autumn; not evergreen.

DENTATE: toothed; the saw-like margin of a leaf.

DISK: an enlargement of a flower receptacle; also the flower head of the tiny center blooms in such flowers as asters and daisies.

DISSECTED: finely divided into numerous segments, as of leaves.

ENTIRE: without divisions, lobes, or toothed margin, as of leaves.

EPIPHYTE: a plant growing upon another plant but not organically connected with it.

EVERGREEN: bearing green leaves at all seasons.

FILAMENT: the stalk of the stamen bearing the anther on top.

FRUIT: the seed-bearing organ of a plant, regardless of its form (a botanical definition).

GLABROUS: hairless, as of leaves.

GLAUCOUS: covered with white or bluish bloom, as of leaves.

GLOBOSE: spherical, globe-like, as a fruit.

HABITAT: the usual environment of a species.

HEAD: a dense cluster of small stemless flowers, such as asters.

HERB: botanically, a plant with no woody stem.

HIRSUTE: coarse, hairy, as of leaves or stems.

HOOD: a modified petal or sepal resembling a monk's hood.

HYBRID: the offspring from the crossing of two species.

INDIGENOUS: native to a region.

INFLATED: bladderlike, as of a calyx or seed container.

INFLORESCENCE: flowering portion of a plant.

LEAFLET: a single division of a compound leaf.

LEGUME: the fruit of the pea family, usually a capsule with two valves splitting along the back.

LIP: a division of a two-lobed flower, usually used in reference to orchids and mints.

LOBED: deeply divided, as of a lobed leaf.

NATURALIZED: not native, but well-established as part of the flora of a region.

NODE: the often thick point on a stem where the leaf or group of leaves is joined to the stem.

OBTUSE: blunted or rounded at the tip, as of leaves.

OPPOSITE: arising on a stem in pairs, as of leaves.

OVARY: the ovule-bearing organ at base of pistil which later develops seeds.

PARASITIC: growing on and deriving nourishment from another plant.

PEDICEL: the stem or support of a single flower.

PELTATE: attached to the stalk at some point on the lower surface, other than the margin, as of leaves.

PERENNIAL: living year after year, in contrast to an annual or biennial.

PERFOLIATE: opposite leaves with clasping bases united, as if pierced by the stem.

PETAL: a division of the corolla, usually colored.

PETIOLE: the leaf stalk.

PINNATE: feathery; segments of a leaf arranged on each side of a common axis.

PISTIL: a flower's seed-producing organ, consisting of ovary, style, and stigma.

PLUMOSE: feather-like, as an organ on a seed.

PROSTRATE: flat on the ground, as the growth habit of a trailing plant.

RACEME: many stalked flowers borne on an elongated axis or stem.

RAY OR RAY FLOWER: The marginal flowers on a flower head, in contrast to the disk flowers in the center, such as daisies.

RECEPTACLE: the enlarged or bulbous end of the flower stalk bearing numerous small flowers, such as asters.

ROOTSTOCK: the usually thickened underground stem, often with small roots at the joints.

RUNNER: a slender trailing branch which roots at intervals.

SAGITTATE: shaped like an arrowhead, as of leaves.

SAPROPHYTE: a plant which grows on dead organic matter.

SCALE: a minute leaf at the base of a shoot or stalk, or the small bracts at flower bases.

SCAPE: a leafless or almost leafless flower stalk arising from the base of a plant.

SEPAL: a division of the calyx, the outer part of a flower, often green but sometimes colored.

SERRATE: a sharp, tooth-like margin, as in some leaves.

SESSILE: stalkless, such as a stemless leaf or flower.

SHEATH: the lower portion of a leaf surrounding the stem, often thin or papery.

SPADIX: a flower spike with a fleshy axis, such as the arum or calla.

SPATHE: a large bract or sheath, often colorful, enclosing a flower spike or spadix, as above.

SPIKE: an elongated compact flower spike, often terminal, with many stemless flowers, such as mullein.

SPUR: a hollow sac-like or tubular extension of a flower, such as in the columbine or many orchids.

STAMEN: the flower organ which produces pollen.

STIGMA: the tip of a pistil through which fertilization by pollen grains is accomplished.

STIPULE: one of a pair of leafy appendages borne at the base of petioles.

STOLON: a trailing basal branch rooting at the joints.

STYLE: the slender stalk of the pistil joining the stigma to the ovary.

SUCCULENT: juicy or fleshy, as of a leaf or a fruit.

TENDRIL: a slender, thread-like, spirally coiled organ of attachment developing from a modified leaf or stem, such as in peas.

TERNATE: arranged in threes.

THROAT: the opening of a tubular flower, such as in honeysuckle.

TOOTHED: marginal serrations, as on a leaf.

TUBER: a short, thickened underground branch, such as a potato.

Bibliography

Appleton's Cyclopedia of American Biography. 1887. Wilson, J. G. and Fisher, John, eds. New York: D. Appleton Co.

Bailey, L. H. 1933. *How Plants Got Their Names.* New York: Macmillan Co.

———— 1947. *Standard Cyclopedia of Horticulture.* New York: Macmillan Co.

Berglund, B. and Bolsby, C. E. 1971. *The Edible Wild.* Toronto: Pagurian Press.

Bigelow, Jacob. 1817. *American Medical Botany.* Boston: Cummings & Hilliard.

Britten, James. 1886. *Dictionary of English Plant Names.* London: Trubner & Co.

Brooklyn Botanic Garden. 1972. *Dye Plants and Dyeing—A Handbook.* Brooklyn, N.Y.: Brooklyn Botanic Garden.

Cassell's New Latin Dictionary. 1968. Simpson, D. P., ed. New York: Funk & Wagnalls Co.

Coats, A. M. 1956. *Flowers and Their Histories.* New York: Pitman Co.

———— 1968. *Garden Shrubs and Their Histories.* New York: E. P. Dutton & Co.

Crowhurst, Adrienne. 1972. *The Weed Cookbook.* New York: Lancer Books.

Dictionary of American Biography. 1928. Johnson, Allen, ed. New York: Charles Scribner's Sons.

Dictionary of National Biography. 1908 et seq. Stephen, Leslie and Lee, Sidney, eds. London: Smith, Elder & Co.

Dictionary of Scientific Biography. 1970. Gillispie, C. C., editor-in-chief. New York: Charles Scribner's Sons.

Earle, John. 1880. *English Plant Names From the 10th to the 15th Centuries.* Oxford: Clarendon Press.

Fernald, M. L. and Kinsey, A. C. 1958. *Edible Wild Plants of Eastern North America*. New York: Harper & Row.

Friend, Hilderic. 1884. *Flowers and Flower Lore*. London: Sonnenschein & Co.

Gibbons, Euell. 1962. *Stalking the Wild Asparagus*. New York: David McKay Co.

———— 1966. *Stalking the Healthful Herbs*. New York: David McKay Co.

Graf, Alfred B. 1968. *Exotica III, Pictorial Cyclopedia of Exotic Plants*. Rutherford, N.J.: Roehrs Co.

Gray's Manual of Botany. 1950. 8th Ed. Fernald, M. L., ed. New York: American Book Co.

Grieve, M. 1971. *Culinary Herbs and Condiments*. New York: Dover Publications.

Hargrave, Basil. 1969. *Origins and Meanings of Popular Phrases and Names*. Detroit: Gale Publishing Co.

Harris, B. C. 1971. *Eat the Weeds*. Barre, Mass.: Barre Publishers.

Healey, B. J. 1972. *Gardener's Guide to Plant Names*. New York: Charles Scribner's Sons.

Index Kewensis. 1895. Jackson, B. D., ed. Oxford: Clarendon Press.

Lehner, E. 1960. *Folklore and Symbolism of Flowers*. New York: Tudor Publishing Co.

National Cyclopedia of American Biography. 1898. New York: James T. White Co.

National Geographic Society. 1924. *Book of Wild Flowers*. Washington, D.C.: National Geographic Society.

New English Dictionary, A. 1888. Murray, J. A. H., ed. Oxford: Clarendon Press.

New York Botanical Garden. 1965. *Biographical Notes Upon Botanists*. Barnhard, J. H., ed. Boston: G. K. Hall & Co.

Northcote, Lady Rosalind. 1971. *Book of Herb Lore*. New York: Dover Publications.

Notable American Women, 1607–1950—A Biographical Dictionary. 1971. James, Edward T., ed. Cambridge: Harvard University Press.

Peterson, R. T. and McKenny, Margaret. 1968. *A

Field Guide to Wild Flowers. Boston: Houghton Mifflin Co.

Plowden, C. C. 1970. *A Manual of Plant Names.* London: George Allen & Unwin.

Prior, R. C. A. 1870. *On the Popular Names of British Plants.* London: Williams & Norgate.

Pritzel, G. A. 1851. *Thesaurus of Literature of Botany.* Leipzig: Brockhaus.

Quinn, Vernon. 1939. *Stories and Legends of Garden Flowers.* New York: F. A. Stokes Co.

Rickett, H. W. 1966. *Wild Flowers of the United States.* New York: McGraw-Hill Book Co.

Rohde, E. S. 1969. *A Garden of Herbs.* New York: Dover Publications.

Royal Horticultural Society. 1965. *Dictionary of Gardening.* Chittenden, F. J., ed. Oxford: Clarendon Press.

Sachs, Julius von. 1906. *History of Botany, 1530–1860.* Oxford: Clarendon Press.

Sanecki, Kay N. 1956. *Wild and Garden Herbs.* New York: Transatlantic Arts.

Skinner, C. M. 1925. *Myths and Legends of Flowers, Trees and Plants.* Philadelphia: J. B. Lippincott & Co.

Smith, A. W. 1963. *A Gardener's Book of Plant Names.* New York: Harper & Row.

Sturtevant's Notes on Edible Plants. 1919. Hedrick, U. P., ed. New York: J. B. Lyon & Co.

Taylor, R. L. 1952. *Plants of Colonial Days.* Williamsburg, Va.: Colonial Williamsburg, Inc.

Thiselton-Dyer, T. F. 1889. *Folklore of Plants.* New York: D. Appleton Co.

Turner, William. 1548. *The Names of Herbes.* James Britten, ed., 1881 edition. London: English Dialect Society.

U.S. Pharmacopeial Convention. 1970. *United States Pharmacopeia.* Easton, Pa.: Mack Publishing Co.

Uphof, J. C. T. 1968. *Dictionary of Economic Plants.* Wurzburg, Germany: J. Cramer.

Van Brunt, E. R. 1943. *Culinary Herbs: Their Culture, Traditions and Uses.* New York: Brooklyn Botanic Garden Record, Vol. 32, No. 1.

Webster's New International Dictionary of the En-

glish Language. 1936. Springfield, Mass.: G. & C. Merriam Co.

Weiner, M. A. 1972. *Earth Medicine—Earth Foods*. New York: Macmillan Co.

Wilkinson, A. E. 1943. *Flower Encyclopedia and Gardener's Guide*. Philadelphia: Blackiston & Co.

Wise Garden Encyclopedia. 1970. Seymour, E. L. D., ed. New York: Grosset & Dunlap.

Wyman, Donald. *Wyman's Gardening Encyclopedia*. 1971. New York: Macmillan Co.

Index

Index

Index

Index

Index

Index

Index

Index

Index

Sedum
  acre, 125
  rosea, 226
  sarmentosum, 255
  telephium, 167
Seedbox, 235
Self-heal. *See* Heal-all
Sempervivum tectorum, 136
Seneca Snakeroot, 236
Senecio
  aureus, 123
  cruetus, 58
  viscosus, 129
Senna, Wild, 236
Sensitive Plant, Wild, 236
Sesame, 236
Sesamum indicum, 236
Sesuvium maritimum, 235
Shallots, 237
Shamrock, 237
Shepherd's Purse, 237
Sherardia arvensis (Sherard, William), 172
Shinleaf. *See* Pyrola
Shoe-black Plant, 225
Shooting Star, 238
Shortia galacifolia (Short, Charles W.), 191
Shreber, Johann C. D., 287
Shrimp Plant, 238
Sickle-pod, 238
Sicyos angulatus, 37
Silene
  cucubalus, 25
  noctiflora, 49
  stellata, 44
Silphium
  integrifolium, 226
  perfoliatum, 146
Silver-rod, 238
Silverweed, 239
Singapore Holly, 239
Sinningia speciosa (Sinning, Wilhelm), 121
Sisymbrium
  altissimum, 272
  officinale, 133
Sisyrinchium mucronatum, 29
Sium
  sisarum, 240
  suave, 198
Skeletonweed, 239
Skimmia, Japanese, 239
Skirret, 240
Skullcap, Mad-dog, 240
Skunk Cabbage, 240
Skyrocket, 240
Sloter, Logan, 77

Smartweed, 241
Smartweed, Swamp, 241
Smilacina racemosa, 102
Smilax
  herbacea, 47
  rotundifolia, 49
Smithiana (Smith, Matilda), 264
Snakemouth. *See* Pogonia, Rose
Snakeplant, 241
Snakeroot, Black, 241
Snakeroot, Seneca, 236
Snakeroot, Virginia. *See* Birthwort
Snakeroot, White, 242
Snapdragon, 242
Snapweed. *See* Impatience
Sneezeweed, Bitter, 242
Sneezeweed, Common, 242
Snowberry, 243
Snowdrop, 243
Snow-on-the-mountain, 243
Soapwort. *See* Bouncingbet
Solanum
  carolinense, 142
  dulcamara, 23
  melongena v. esculentum, 93
  nigrum, 189
  pseudo-capsicum, 151
  tuberosum, 212
Soleirol, J. F., 13
Solidago
  bicolor, 238
  odora, 124
Solomon's Lily. *See* Black Calla
Solomon's Seal, 243
Solomon's Seal, False, 102
Sonchus oleraceus, 245
Sorrel, Sheep, 244
Sorrel, Wood, 195, 293
Southernwood, 244
Sow Thistle, 245
Soybean, 245
Spanish Needles, 245
Sparganium americanum, 39
Spatterdock, 245
Spatterdock, Small, 246
Spearmint, 246
Spearwort, 246
Specularia perfoliata, 278
Speedwell, Common, 246
Speedwell, Purslane, 247
Speedwell, Water, 33
Spergula arvensis, 250
Spergularia rubra, 230, 250
Spiderflower, 247
Spider Lily, 247
Spider-lily, Inland, 247
Spiderwort, 248
Spikenard, 248

Index

Spinach, 248
Spinach, New Zealand, 248
Spinacia oleracea, 248
Spiraea
  latifolia, 177
  tomentosa, 253
  van houttei, 32
Spotted Knapweed. *See* Star Thistle
Spring Beauty, 249
Spurge, Cypress, 249
Spurge, Flowering, 249
Spurge, Japanese (Pachysandra), 197
Spurge, Laurel, 249
Spurge, Leafy, 250
Spurrey, Corn, 250
Spurrey, Sand, 250
Squash, Crookneck, 250
Squash, Winter, 251
Squawroot, 251
Squill, 251
Squirrel Corn, 251
Stachys
  palustris, 133
  tenuifolia, 134
Staphylea trifolia, 25
Starflower, 251
Star Grass. *See* Colicroot
Star Grass, Yellow, 251
Star-of-Bethlehem, 252
Star Thistle, 252
Star Thistle, Yellow, 252
Star Tulip. *See* Mariposa Lily
Statice. *See* Sea Lavender
Statice, Russian, 253
Steeplebush, 253
Steironema ciliatum, 168
Stellaria
  graminea, 253
  media, 54
Stenanthium gramineum, 103
Sternbergia lutea (Sternberg, Kaspar M. Von), 98
Stickseed, 253
Sticktight. *See* Beggarticks
Stitchwort, 253
Stock, Brompton, 254
Stock, Night-scented, 254
Stokesia laevis (Stokes, Dr. Jonathan), 254
Stonecress, 254
Stonecrop. *See* Goldmoss
Stonecrop, Mountain, 255
Stonecrop, Stringy, 255
Storksbill, 255
Strawberry, False, 102
Strawberry, Wild, 255

Strawberry Bush. *See* Burning Bush
Strawberry Geranium, 255
Strawflower, 256
Streptocarpus rexii, 46
Streptopus roseus, 274
Strophostyles umbellata, 17
Stylosanthes biflora, 202
Suaeda maritima, 233
Summer Cypress, 155
Sundew, Roundleaf, 256
Sundew, Thread-leaved, 256
Sundrops, 257
Sunflower, Common, 257
Sunflower, False. *See* Heliopsis
Sunflower, Mexican, 257
Sun Rose, 258
Swallowwort, Black, 258
Swampcandle, 258
Sweet Basil, 258
Sweetbriar, 259
Sweet Cicely. *See* Cicely, Sweet
Sweet Clover, White, 259
Sweet Clover, Yellow, 259
Sweet Coltsfoot, 259
Sweet Fennel. *See* Fennel, Florence
Sweet Flag, 260
Sweet Gale, 260
Sweet Olive, 260
Sweet Pea, 201
Sweet Potato, 260
Sweet Rocket. *See* Dame's Rocket
Sweetshrub, 261
Sweet William, Garden, 261
Sweet William, Wild, 261
Swiss Chard, 53
Swiss-cheese Plant, 261
Symphoricarpus
  albus, 243
  orbiculatus, 66
Symphyton officinale, 65
Symplocarpus foetidus, 240
Syngonium podophyllum, 262
Syringa vulgaris, 164

Tabasco Pepper. *See* Christmas Pepper
Taenidia integerrima, 207
Tagetes erecta, 173
Talinum teretifolium, 102
Tamarisk, Kashgar, 263
Tamarix hispida, 263
Tanacetum vulgare, 263
Tangleberry. *See* Dangleberry
Tansy, Common, 263
Tape Grass. *See* Eelgrass
Taraxacum, 80

326

Index

Xerophyllum asphodeloides, 273
Xyris caroliniana, 296

Yam. *See* Sweet Potato
Yam, Chinese, 295
Yarrow, 295
Yellow Cress, 296
Yellow-eye Grass, 296
Yellow Flag, 296
Yucca filamentosa, 296

Zantedeschia aethiopica (Zantedeschi, Francesco), 43
Zea mays, 68
Zebra Plant, 297
Zenobia, Dusty (Zenobia Septimia), 297
Zephyranthes atamasco, 297
Zephyr-lily, 297
Zespedes, V. M. de, 39
Zinnia elegans, 298
Zizia aurea (Ziz, I. B.), 123